Make:

Getting Started with Soldering

A Hands-On Guide to Making Electrical and Mechanical Connections

Marc de Vinck

Maker Media, Inc.
San Francisco

Printed in Canada.

Published by Maker Media, Inc., 1700 Montgomery Street, Suite 240, San Francisco, CA 94111

Maker Media books may be purchased for educational, business, or sales promotional use. Online editions are also available for most titles (*safaribooksonline.com*). For more information, contact our corporate/institutional sales department: 800-998-9938 or *corporate@oreilly.com*.

Publisher: Roger Stewart
Editor: Roger Stewart
Copy Editor: Elizabeth Campbell, Happenstance Type-O-Rama
Proofreader: Scout Festa, Happenstance Type-O-Rama
Interior Designer and Compositor: Maureen Forys, Happenstance Type-O-Rama
Cover Designer: Maureen Forys, Happenstance Type-O-Rama
Indexer: Valerie Perry, Happenstance Type-O-Rama

November 2017: First Edition

Revision History for the First Edition

2017-11-03 First Release

See *oreilly.com/catalog/errata.csp?isbn=9781680453843* for release details.

978-1-680-45384-3

Safari® Books Online

Safari Books Online is an on-demand digital library that delivers expert content in both book and video form from the world's leading authors in technology and business. Technology professionals, software developers, web designers, and business and creative professionals use Safari Books Online as their primary resource for research, problem solving, learning, and certification training. Safari Books Online offers a range of plans and pricing for enterprise, government, education, and individuals. Members have access to thousands of books, training videos, and prepublication manuscripts in one fully searchable database from publishers like O'Reilly Media, Prentice Hall Professional, Addison-Wesley Professional, Microsoft Press, Sams, Que, Peachpit Press, Focal Press, Cisco Press, John Wiley & Sons, Syngress, Morgan Kaufmann, IBM Redbooks, Packt, Adobe Press, FT Press, Apress, Manning, New Riders, McGraw-Hill, Jones & Bartlett, Course Technology, and hundreds more. For more information about Safari Books Online, please visit us online.

How to Contact Us

Please address comments and questions to the publisher:

Maker Media
1700 Montgomery St.
Suite 240
San Francisco, CA 94111

You can send comments and questions to us by email at *books@makermedia.com*.

Maker Media unites, inspires, informs, and entertains a growing community of resourceful people who undertake amazing projects in their backyards, basements, and garages. Maker Media celebrates your right to tweak, hack, and bend any Technology to your will. The Maker Media audience continues to be a growing culture and community that believes in bettering ourselves, our environment, our educational system—our entire world. This is much more than an audience, it's a worldwide movement that Maker Media is leading. We call it the Maker Movement.

To learn more about Make: visit us at *makezine.com*. You can learn more about the company at the following websites:

Maker Media: *makermedia.com*
Maker Faire: *makerfaire.com*
Maker Shed: *makershed.com*
Maker Share: *makershare.com*

To my wife, Christa.

Contents

Foreword

The world of electronics is waiting for you—a fantastic land of LEDs, buttons, motors, and microcontrollers. And what is your key to this kingdom? The common soldering iron! This humble item has been tossed into toolboxes for years, to be pulled out whenever a household wire is loose. Whether your soldering iron is a pen-type, battery- or butane-powered, or a fancy stand model, you will be able to do **a lot** with soldering.

First of all, you can fix things around you!

How many times have you had to toss out a really nice pair of headphones because of a worn-out wire? Now, that will be a 10-minute job. Or maybe your bicycle headlamp goes on the fritz whenever you jump the curb? Get in there with your soldering iron and fix it—good as new.

And then, you can make new things!

Upgrade your costume, cosplay, or festival attire with LEDs, soldered in place. If you do theater A/V, your ability to make custom cables will be very handy. Or pick up a kit, and get into the relaxing rhythm of "cross-stitching with molten metal" (as I like to call it).

Soldering may seem a little intimidating at first—it's hot, it's metal, what kind of solder should I use? But it's not that tough; millions of people have picked up this handy skill. And Marc here is an expert—not just at soldering, but at knowing how to get **you**, the beginner, going!

Are you ready? Then let's get started . . .

Limor "Ladyada" Fried, Adafruit

Preface

My adventures in soldering began with a kit that I ordered from a person in Long Island City about 10 years ago. The kit was a mini persistence-of-vision (POV) circuit. It consisted of a series of LEDs that would turn on and off at a very specific rate, and if everything went well, you could wave it in the air to spell out a word or create a graphic. It was magical!

The only problem was it had to be soldered. Although I had soldered before, I had never soldered anything as ambitious as a printed circuit board. I ended up purchasing a very inexpensive soldering iron and somehow managed to solder it together and get it to work. The feeling of accomplishment was amazing, and I was hooked on learning more about soldering electronics.

Learning to solder was a bit intimidating. There were all types of irons and tips, and even different types of solder. Who knew there would be so many different choices?

By the way, that cheap iron I bought for the POV project subsequently caught fire! (Don't worry! We'll also talk about how to make sure you choose the right equipment!)

I learned almost everything I know about soldering by getting advice from other Makers, looking up information and videos online, and, most of all, from years of practice and first-hand experience soldering and making electronic kits. Along the way, I designed and helped launch the Learn to Solder program at Maker Faire, which is now sponsored by Google and has taught thousands of young people this valuable skill.

In this book, I will give you helpful advice on navigating the available tools and materials as to what may be your best buying options. I'll also show the basics of soldering through-hole components, and then later on, more advanced techniques about soldering surface-mount components that are typically reserved for robots and professionals. I hope this book also helps you avoid some common pitfalls, or at least minimize their impact. I'll show you examples of the typical things that can (and will) go wrong when soldering, and I'll also show you how to fix them.

In full disclosure, I should mention that many of the products used in this book were generously donated by Adafruit Industries (*adafruit.com*). I'm grateful for their help. Few companies offer electronics enthusiasts such great products and customer support. If you want to learn more about electronics, you should check out their website—it's more than just a store. They also have an excellent learning system (*learn.adafruit.com*). You might even see a few of my projects there!

That said, you do have options when shopping for soldering irons and supplies. Other online sites that cater to electronics hobbyists include Sparkfun (*sparkfun.com*), Jameco (*jameco.com*), Digi-Key (*digikey.com*), and many others in the United States and around the world. You can also typically find a basic soldering iron and related supplies at local hobby shops.

Oh, and as it turns out, that POV kit I mentioned back at the beginning of this preface was created by a woman named Limor Fried. Thanks, Limor, for your continued support and for being the catalyst for my own Making adventures!

1

What Is Soldering?

What exactly *is* soldering? Fortunately, we can probably agree on what it is a lot more easily than we can agree on how to pronounce it! Is it sah-*der-ing* or soul-*der-ing*? How you pronounce it depends largely on where you live, so the debate will continue—but, really, how you say it is up to you!

No matter how you pronounce *soldering*, in its most basic form it is a mechanical connection between two or more similar or dissimilar materials made by heating and melting a third "filler" material (see Figure 1-1). The two materials you're joining do not melt during the process of soldering—only the solder melts. It's a fairly simple process that has been around for at least five thousand years. Yes, that's right . . . five *thousand* years!

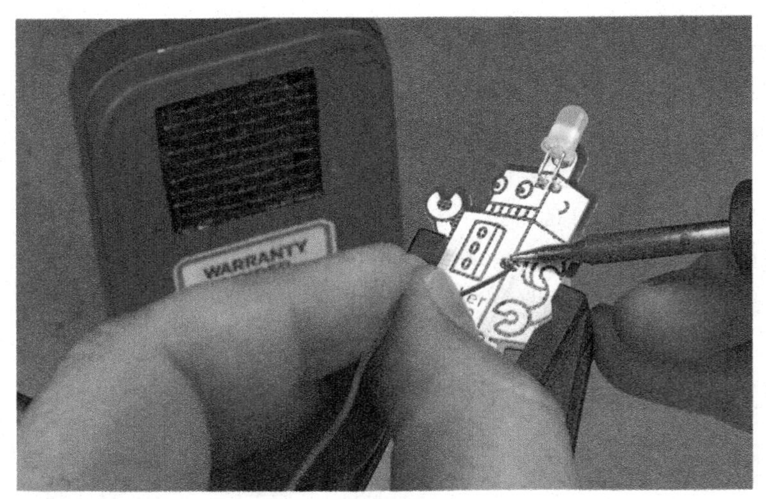

FIGURE 1-1: Depositing solder on a "learn to solder" pin designed by the author

There are dozens of ways to solder. The fundamental process is always the same, but how you make the mechanical connection can vary a lot. You can do so with chemicals or heat—and innovators keep coming up with new ways to do both.

The two most common methods of generating heat for soldering are using a flame or using an electrical source. The electrical source can generate its heat by radiation, induction, conduction, arcing, or through several other methods.

Generating a flame from a gas source is the typical chemical process, but it has variations, too. In this book, we are going to focus on using electrical heat generated by a soldering iron, as shown in Figure 1-2.

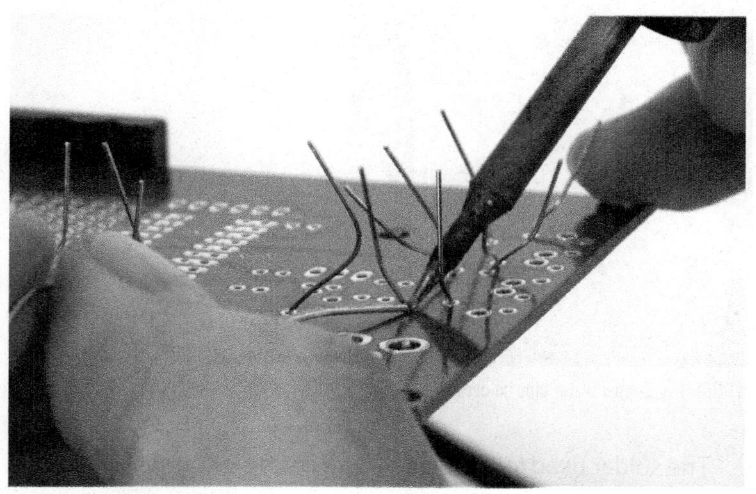

FIGURE 1-2: The business end of a soldering iron in action

OTHER TYPES OF SOLDERING

Your first experience with soldering might have been plumbing. Most homes have copper water pipes, and as in Figure 1-3, they are joined with solder. The metal-based solder and chemical flux (which prevents oxidation) used on the copper pipes is different from the type used in electronics, but the process is generally the same. The pipes are cleaned, and a flux is applied to keep the metal from oxidizing. Heat is applied, typically with a flame from a blow torch, and solder is then introduced into the joint. This type of soldering uses lower temperatures than other methods, and is sometimes referred to as *soft soldering*.

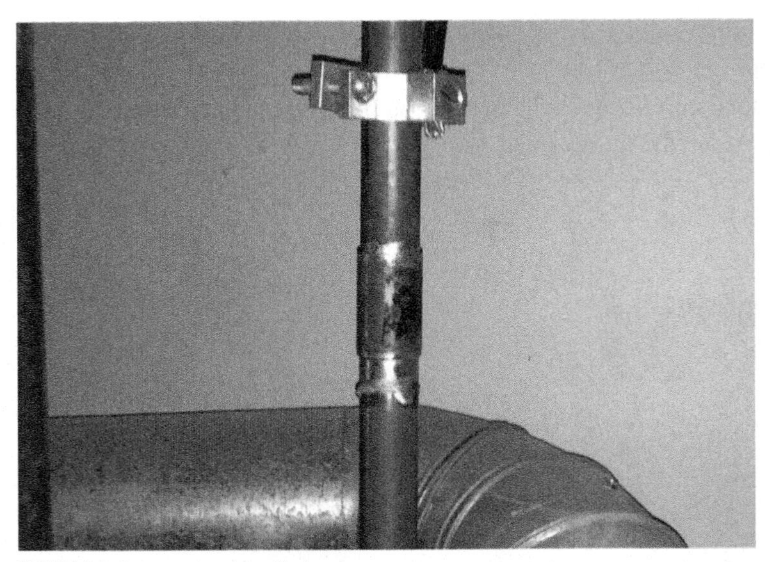

FIGURE 1-3: Copper water pipe soldering

The solder used for copper pipes must be lead-free, since lead is toxic if consumed, and the flux must be safe for use on water pipes. Once the soldered connection has cooled, the pipes are ready to be used, often for years without fail.

Silver soldering (see Figure 1-4) is another common soldering technique. It's also called *hard soldering* because of the higher temperatures (above 840°F or 450°C) used to melt the filler materials. This is opposed to *soft soldering* (below 840°F or 450°C) used for electronics and copper pipes.

Silver soldering is the first type of soldering that I was taught, and I hope to find time to return to it again someday, as I really enjoyed the process. (Another example of my work with silver soldering is shown in Figure 1-5.) The silver and steel spring-loaded box shown in Figure 1-4 is an item I made with that process many years ago.

FIGURE 1-4: A spring-loaded box created with silver soldering made by the author

Hard soldering is the most common type of soldering used for jewelry-making and repair, but it has other uses, too. Hard soldering has one main advantage over soft soldering: it forms a much stronger bond between the materials. The downside is that the much higher temperatures make it impractical for temperature-sensitive electronic components.

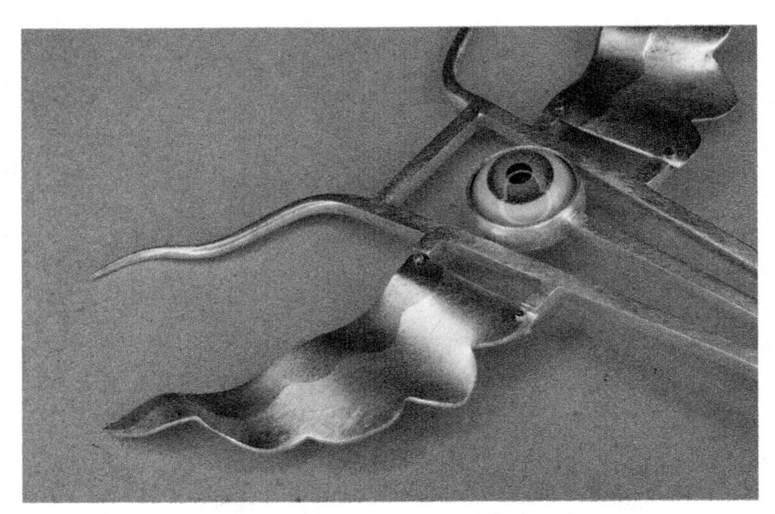

FIGURE 1-5: Detail of copper and silver sculptural spoon made by the author

Brazing, as seen in Figure 1-6, is another type of hard soldering that utilizes a high temperature to join materials. With brazing, the filler material is typically a brass alloy that is heated to just above its melting point and, through capillary action, is drawn into the joint between the materials to be connected. It makes an incredibly strong joint, but not typically as strong as a well-executed welded joint.

THE DIFFERENCE BETWEEN SOLDERING AND WELDING

Welding is not soldering. With welding, the materials to be joined are *also* melted in the process of making the connection. You typically still use a filler material when welding, but the key to a good weld is the fusion of the materials being welded. A weld joint should be as strong as the base materials being welded. (See Figure 1-7.) Just as with soldering, there are dozens of ways to weld materials together.

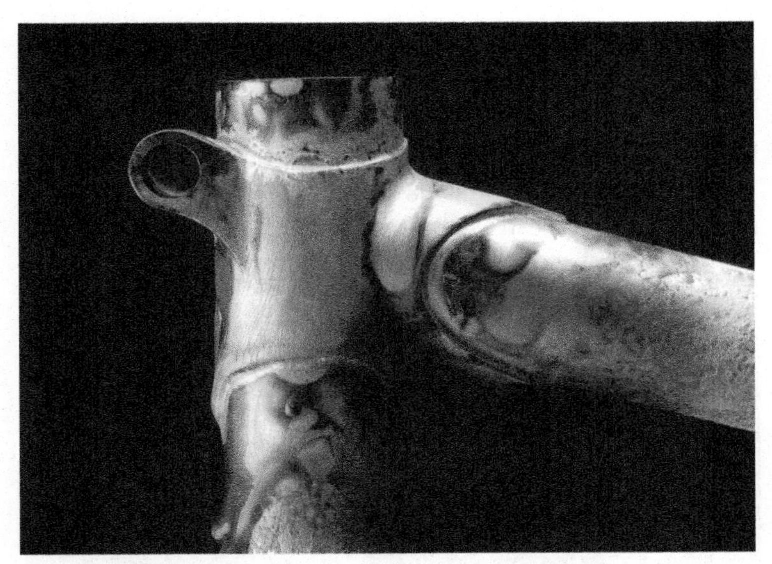

FIGURE 1-6: Brazed seat lug from "Building a frame" by Phil Gradwell IMAGE COURTESY OF FLICKER COMMONS, LICENSED UNDER CC BY 2.00

FIGURE 1-7: A welded connector for an off-road vehicle IMAGE COURTESY OF ROCK BUG BUILD ALBUM, FLICKR COMMONS, LICENSED UNDER CC BY 2.0

The most common welding technique is to use a flame or electrical arc (Figure 1-8), but you can also weld with friction, a laser, and even ultrasonic waves. Note that I have been using the term *materials* and not *metal*. That's because you can weld more than just metal. Plastic welding is a very common technique, and it's something you can even do at home with a few basic tools.

FIGURE 1-8: A Naval hull technician performs TIG welding on a steel shelf IMAGE COURTESY OF OF LPD-18 PHOTO GALLERY BY NAVAL SURFACE TECHNICIANS, FLICKR COMMONS, LICENSED UNDER CC BY 2.0

SOLDERING ELECTRONICS

We've looked at several types of soldering, and we've examined how welding and soldering are different. But this book is about soldering hobby electronics, like those in Figure 1-9, and that is what we are going to focus on for the rest of the book.

The most basic definition of soldering that I emphasize with my students is:

Soldering is an electrical and *mechanical connection.*

FIGURE 1-9: Electronic kit waiting to be soldered

It's important to understand that soldering electronics is not just about making an electrical connection. Understanding the mechanical component of soldering is just as critical. But before we delve any further into the art and science of soldering, we will need to talk about the tools and materials you'll be working with. That is the subject of Chapter 2, "Basic Soldering Tools and Materials."

2

Basic Soldering Tools and Materials

There are a lot of tools needed to learn all the different types of soldering skills that I will be teaching you in this book. However, if you are just getting started, you really don't need a lot of tools or equipment—just the basics—and that's where I will begin this chapter. As we progress through the book, and learn new techniques, I will describe all the required additional tools and equipment.

Getting Started

To get started soldering you only need a few basic tools and supplies:

1. Safety glasses
2. Soldering iron with stand
3. Wire cutters
4. Damp sponge
5. Solder

This is the bare minimum of the tools you will need to learn to solder. You will most certainly want a few additional tools and supplies to make your learning experience easier, but you can get away with only these items if you have a very tight budget.

Most of you would guess that a soldering iron is the most important tool you need, but although it's important, it's not the *most* important. Safety comes first, and a comfortable pair of safety glasses is the most important tool (see Figure 2-1). Most people I

FIGURE 2-1: Always wear your safety glasses when soldering!

see soldering do not wear safety glasses—that is, until they have a close call or, even worse, experience some eye damage.

There are some good reasons for wearing safety glasses, most of which involve having two eyes that need to last you the rest of your life. When soldering, the molten solder can sometimes pop and spatter. It's not a large amount of solder, and it doesn't happen often, but when it does, one little blob can do a lot of damage to your eye.

You may face another danger when you trim the wire leads to components, which you inevitably have to do when you finish soldering a connection. Everyone who solders, regardless of their propensity to wear safety glasses, has experienced a scare when it comes to trimming these leads, as seen in Figure 2-2, which have a tendency to go flying around like little needle-like projectiles.

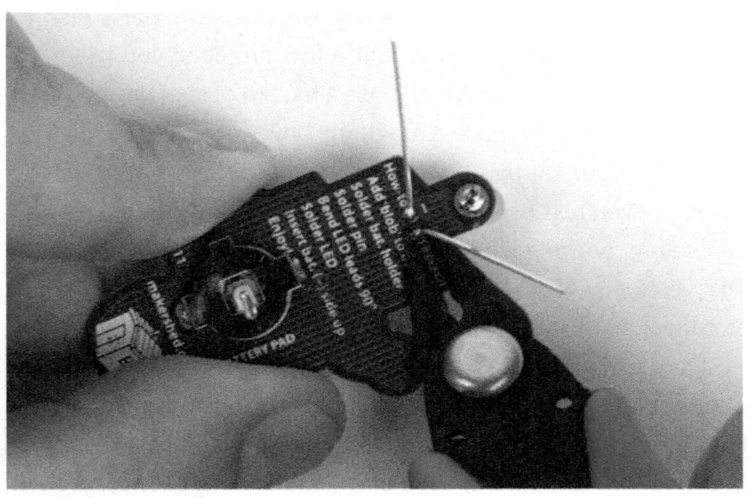

FIGURE 2-2: Trimming wire leads after soldering

Since we are on the topic of clipping leads, let's talk about a good strategy for cutting and collecting them. Always place your hand over a piece of wire that you are cutting, which will shield your face from the wire being trimmed (see Figure 2-3).

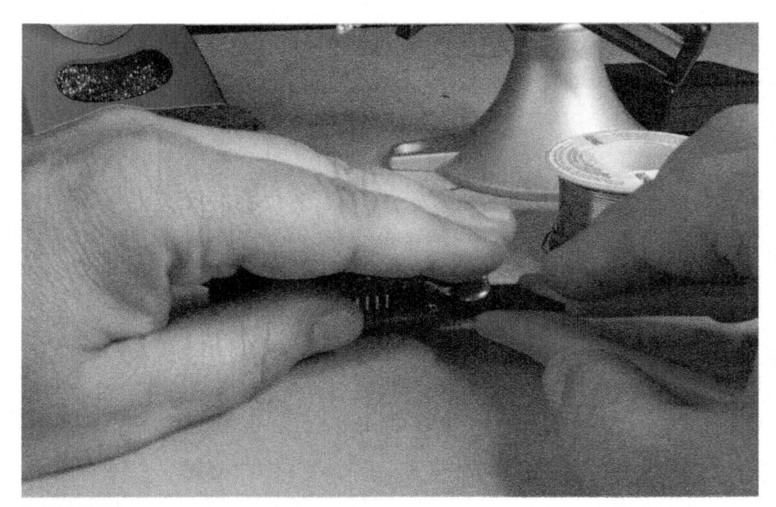

FIGURE 2-3: Hand over lead

It's also a good idea to have a small, magnetic parts dish to store all the wire leads. After just a few projects, you will have hundreds of leads, all of which pose a hazard of shorting out your project. You can purchase a magnetic bowl for just a few dollars, make one yourself by attaching a magnet to any nonmagnetic bowl, or even just use the magnet on its own. In Figure 2-4, you can see a commercial magnetic parts bowl on the left, and my DIY version, which is simply a magnet that I keep on my bench to collect all the trimmed leads.

Now that we have some basic safety out of the way, let's talk about soldering irons. Your soldering iron is the single biggest key to successful soldering. A bad soldering iron usually leads to bad soldering and, therefore, a bad soldering experience. A good soldering iron typically leads to a good soldering experience. It's one of those situations where the quality of the tool is important. The good news is that you can get a good quality soldering iron for not too much money. Figure 2-5 shows my personal soldering iron. I bought it off eBay for about $150 USD. This is not what I

FIGURE 2-4: Magnetic bowl and DIY magnet version

recommend you purchase as your first soldering iron. Don't get me wrong; it's awesome, but it also has some limiting factors for beginners. In fact, I often use a far less expensive iron for soldering due to the ease of changing the temperature, among other things, which I will discuss in greater detail later in this chapter.

FIGURE 2-5: My everyday, professional-quality soldering iron

SO, WHAT SOLDERING IRON SHOULD YOU PURCHASE?

It depends on a few factors, and once you know about them, you can make the best choice for your budget and skill set. Before we talk about your budget, though, there are a few things to consider first.

Comfort is an important factor to consider when you're purchasing an iron. Although it's not always possible to test your soldering iron before you purchase it, you can think about what shape you might prefer.

There isn't a lot of variation in soldering iron ergonomics, but there are some styles with pencil-like handles and others that have a handle with a much larger diameter (see Figure 2-6). In the end, it won't matter too much, but it's something to think about when making your purchase. I have used very inexpensive irons, which typically have a large-diameter handle. They just didn't feel right and made soldering difficult.

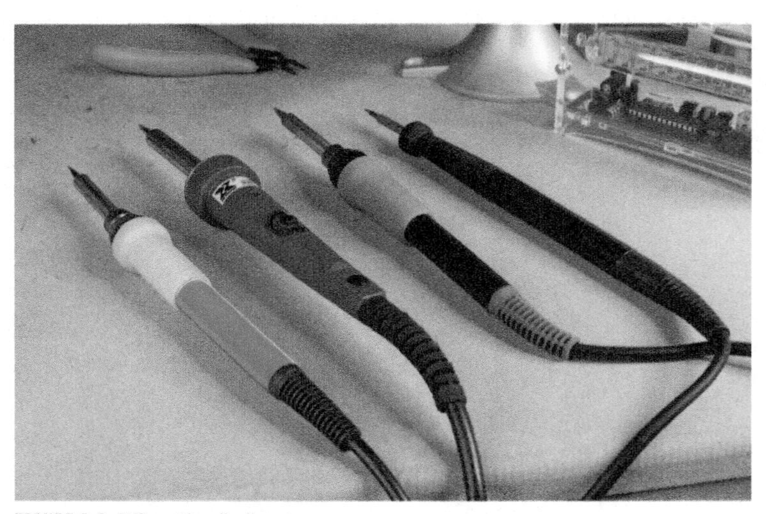

FIGURE 2-6: Different handle diameters

The most important factor (technically two factors) in purchasing your first soldering iron is the wattage of the iron and the ability to vary the wattage. Most irons range from 20 W to 60 W or more. The inexpensive ones do not allow you to adjust the wattage, which adjusts the temperature, nor do they have enough wattage to make soldering fun.

Now you might be asking why wattage is related to fun. It's fairly simple: if your soldering iron has a low wattage, say 15 W, it takes a long time to reach the temperature required to solder a connection. In addition to the heat-up time, it will take longer to recover from soldering the connection, which makes the whole process slow and frustrating. Imagine you want to solder 30 connections on a printed circuit board (PCB). If you use a low-wattage iron, it could take a few minutes to warm up, and 10–20 seconds to recover from each connection you solder. Your 30 connections could take a total of 3–5 minutes to make with a low-wattage iron. Trust me: once you learn to solder, each connection should only take about 3 seconds, and waiting an additional 10 seconds gets frustrating very quickly.

You also need the ability to adjust the wattage or temperature of your iron (see Figure 2-7). Some people like to solder quickly, and some like to solder a little slower. Also, different types of solder require different temperatures, so being able to control the temperature is critical to having a great soldering experience. Fortunately, single-wattage soldering irons are becoming rarer; most have some adjustability.

The Good, the Great, and the Never!

Let's start with the soldering iron that you should *never* buy. I like to refer to these as *beginner irons*, or more commonly, *fire starters* or *fire sticks*. (See Figure 2-8.) They typically cost less than $15 USD, and don't have a decent stand, auto-shutoff, temperature control, or any type of UL (Underwriters Laboratory) listing information. In fact, I have seen these types of irons sold at flea markets with no information at all on the packaging—not even a brand or company name.

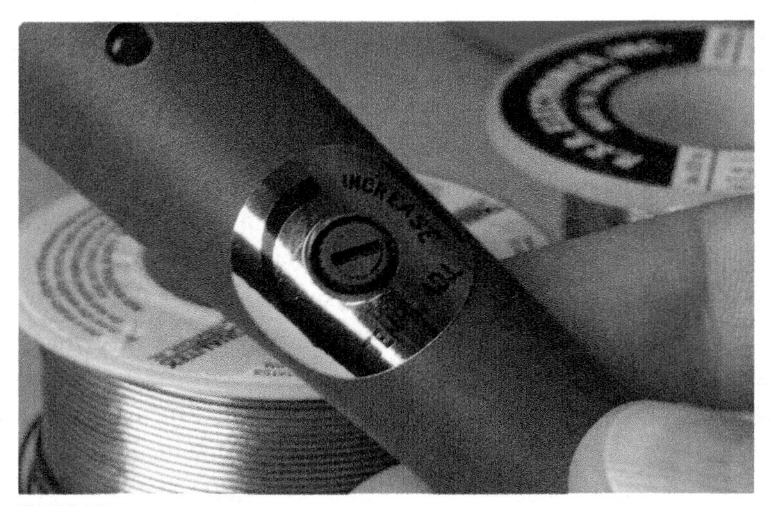

FIGURE 2-7: Temperature variation close-up shot

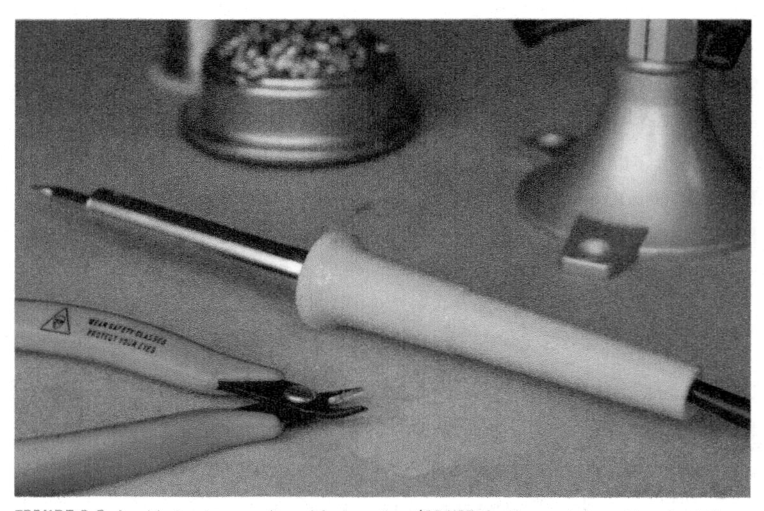

FIGURE 2-8: A soldering iron purchased for less than $10 USD that has no information about the wattage, manufacturer, or UL listing

In addition to the dangers of a non-UL-listed iron, look at the flimsy stand it typically comes with, an example of which is shown in Figure 2-11. Is that what you want to hold up a 700°F soldering iron? Not me, and not in my studio! Now that we're clear about that, let's look at some better choices.

You should think about what your long-term expectations are before you make your purchase. If you just need to solder something occasionally, and want it done well, a simple variable-control, pencil-style soldering iron is perfect.

There are a few things to note about even a simple and inexpensive soldering iron. First, make sure the wattage varies from about 20 W to 50 W. This will give you a nice range of temperatures that will work for 99.9% of your soldering needs. Second, make sure it is certified by UL. Your soldering iron gets hot; 700°F (370°C) or more. Do you really want an iron that poses an electrical hazard? Me neither! The soldering iron in Figure 2-9 meets all the criteria mentioned above and is still inexpensive.

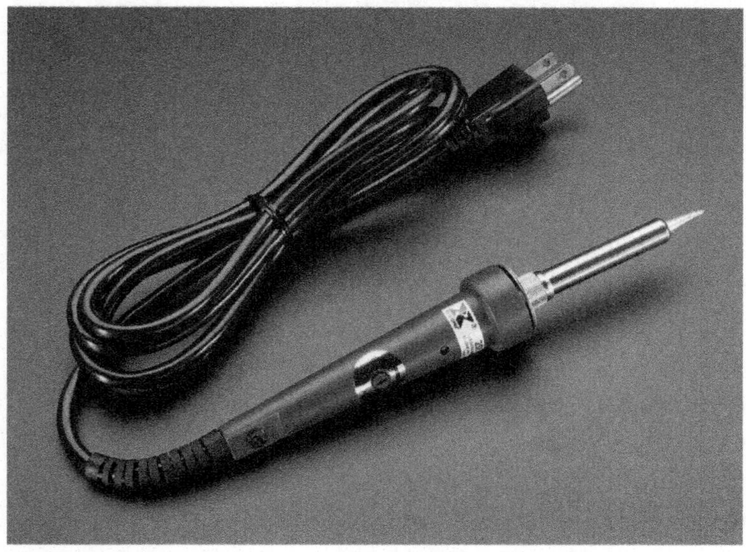

FIGURE 2-9: A 20 W, variable-temperature soldering iron

On the other hand, if you think that you will use a soldering iron more regularly, a proper soldering station is most likely a better choice. They typically heat up and recover faster, have more flexibility in their temperature range, and have a better soldering iron holder. Another nice thing about purchasing a soldering station is that the soldering iron is usually smaller, since the temperature control and other electronics can be housed in the base unit. I really like the Hakko soldering station—the one pictured in Figure 2-10 is from Adafruit Industries. It's a great iron that will last you a very long time, if not a lifetime.

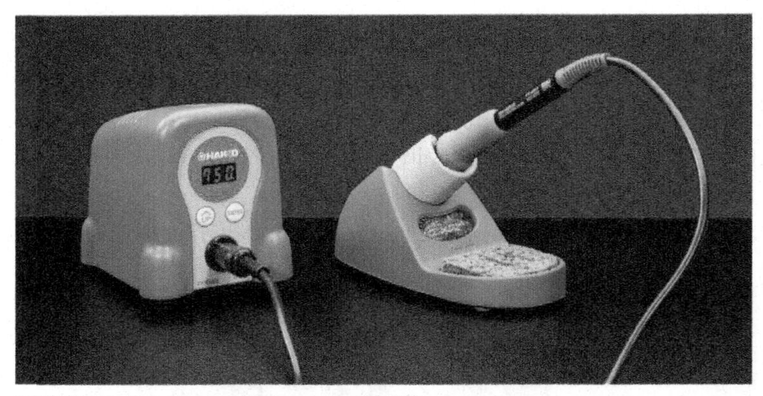

FIGURE 2-10: Digital Hakko soldering station FX-888D IMAGE COURTESY OF ADAFRUIT INDUSTRIES, CC BY-NC-SA 2.0.

As you solder more, then you might consider a professional-grade soldering station. These types of soldering stations have incredibly fast heat-up times—some take just a second or two—and can be used all day, every day, without any problems. The reason I don't recommend buying a professional iron to begin with is mostly due to the fact that you may not need one next week, or next month. What if you decide soldering isn't for you? OK, that's not going to happen! Another reason is that the tips are usually more expensive, and when you start learning you will inevitably ruin a tip. Besides, even though I have a professional, expensive iron, I use my Hakko iron all the time at another station in my lab.

And to be quite honest, I don't really see much of a difference! An avid enthusiast with two soldering stations is not uncommon.

But wait; there's more!

You should seriously consider a few accessories that go hand in hand with your soldering iron. The first one is a proper stand for your iron. It's an important component of your soldering lab that can sometimes be overlooked.

Most, if not all, soldering stations come with a decent stand for your iron, but if you purchase an inexpensive stand-alone soldering iron, it may come with a sheet metal stand, like the one pictured in Figure 2-11, or some kind of cheap, little, bent piece of wire attached to a plastic disk. If so, *do not* under any circumstances use this as your soldering iron stand. Remember: the iron can heat up to over 700°F. Do you want a cheap, flimsy, poorly made, sheet-metal stand to be your only protection against a fire if the iron falls off and touches a wooden surface or other material? How about when you bump the iron, ever so lightly, and it rolls off the stand and burns you!? Just the weight of the cord can be enough to make the iron fall off a flimsy stand. Do yourself a favor and don't even unpack this type of stand—just toss it, or use the metal for another project!

FIGURE 2-11: An example of a soldering iron stand that you should NOT use!

There are much better stands available, and a nice solid stand is a good investment. It also happens to be a fairly inexpensive one, too! Most soldering stations that cost approximately $50 USD or more will come with a decent stand. The less-expensive kits will typically include a stand that looks like a wound-up spring. These work great and are inexpensive. If you purchase a more expensive soldering station, it will most likely come with a more robust version, as pictured in Figure 2-12. All these stands are really good, and will last you a lifetime.

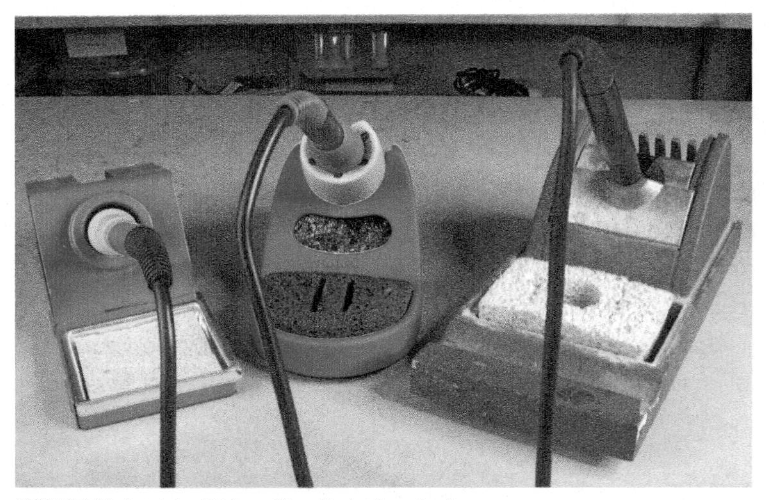

FIGURE 2-12: A variety of high-quality soldering iron stands

But What about the Other Types of Soldering Irons?

There are also pistol-grip-style soldering irons, one of which is shown in Figure 2-13. These are the ones I usually hear about from Makers who found one in the garage and tried to solder their first printed circuit board. It's rarely a happy story, and usually ends in catastrophic failure. These types of soldering irons are only appropriate for larger wire diameters, metals like the ones used for automobile batteries, or even soldering leaded windows. A pistol-grip iron is not the type you want to use for hobby electronics. It's typically difficult to hold, lacks the precision needed for soldering components in electrical circuits, and is higher wattage than necessary. This type is just not appropriate for soldering hobby electronics.

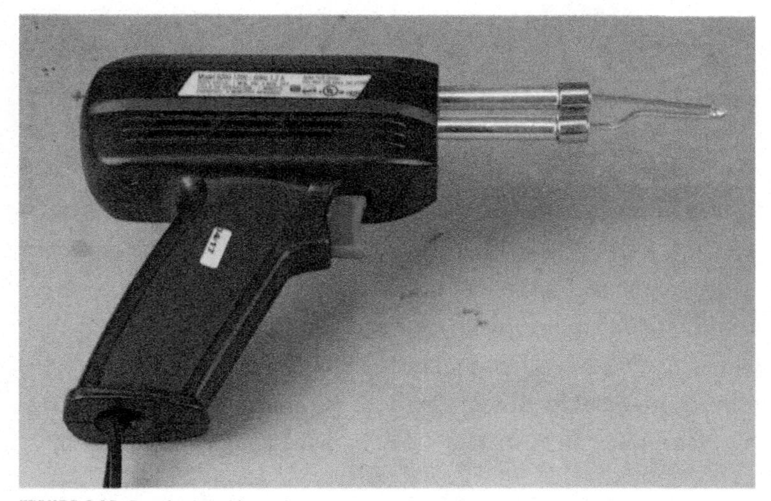

FIGURE 2-13: Pistol-grip soldering iron

Another common type of soldering iron I hear about from hobbyists is the kind powered by gas; typically, butane (shown in Figure 2-14). Again, this type is not ideal for a few reasons. First, you will go through a lot of fuel, which is certainly less convenient than plugging in a standard soldering iron. The other, more important reason is that regulating the temperature is extremely difficult. The solder you use has a specific melting point, and printed circuit boards, along with integrated circuits, are not tolerant to the high heat levels these irons produce.

FIGURE 2-14: Gas-fueled soldering iron

That being said, there are certainly times when a gas-powered iron is a great tool to have on hand. You can't exactly fix a bad wire in a remote location that doesn't have electricity. These mini-torches are great for other tasks, too; they make a great addition to your lab—just don't use it for soldering unless it's 100% necessary.

The Electrical Discharge Soldering Iron

The last type of iron I want to discuss is the electrical discharge soldering iron. These irons work by emitting a quick electrical charge across the tip, which melts the solder or even welds the component in place. The problem with using this type of iron for hobby electronics is the current running through the tip. This current could destroy any sensitive component in your project. As with some of the other less-than-ideal irons I have discussed, there is a time and a place for this type of soldering iron; for example, if you only needed to make a connection between a few light gauge wires at a remote location. However, I would still use an iron powered by butane or other gas source, since it has other uses besides just soldering.

TIP Not all soldering irons have an auto shutoff, and you might not always remember to turn the iron off after you use it. Many years ago, I came up with a simple system to make sure I never leave my iron on for too long: purchase an electrical timer and set the timer to one hour (as shown in Figure 2-15). Plug your iron in, and get to work. If you forget to turn your iron off, the timer will turn it off after one hour. And yes, it sometimes turns off when I'm working, but it's a small price to pay for some extra safety. And remember: the time doesn't matter. Just spin the dial and turn it back on—it's only once an hour.

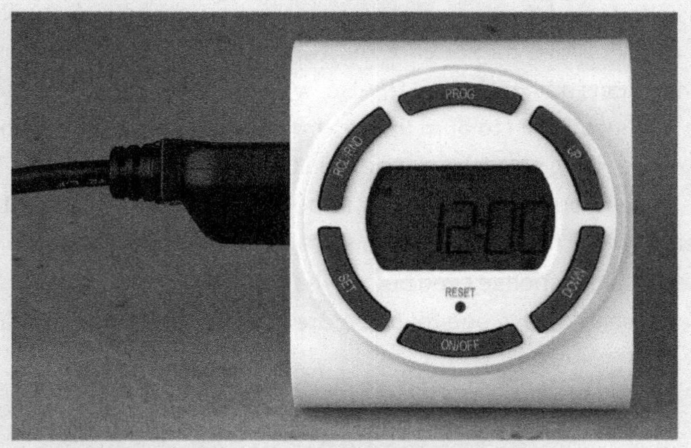

FIGURE 2-15: Soldering iron plugged into a standard electrical timer

Summary of Soldering Irons

Selecting a soldering iron isn't too difficult. If you are having trouble deciding which one to buy, just buy a UL-listed iron with temperature control that fits into your budget. If it costs less than a fast food dinner, save up for better one (but remember that it doesn't have to cost a lot to start learning how to solder).

SOLDERING IRON ACCESSORIES

One other thing to consider when selecting your soldering iron is the availability of replacement tips, including those in different sizes (see Figure 2-16). In general, a medium-size, pointed tip will meet 95% or more of your soldering needs, but the ability to swap out the standard tip for a larger one can be very helpful. The tip of your soldering iron should match the size of the connection being made. A larger component, like a USB port or power jack, requires more heat to make the mechanical connection. A larger tip will make this easier, and the same goes for smaller components. A smaller tip usually allows for better dexterity, which helps you make those connections a bit more easily. Also, when you first learn to solder, it's not uncommon to burn up a tip from improper care of the iron. So again, being able to replace an old tip is a nice feature.

When you're soldering, it's important that you keep the tip clean, which means you will need a soldering sponge. You will need a solder sponge even before you make your first connection, so you should have one ready before you even plug your iron in.

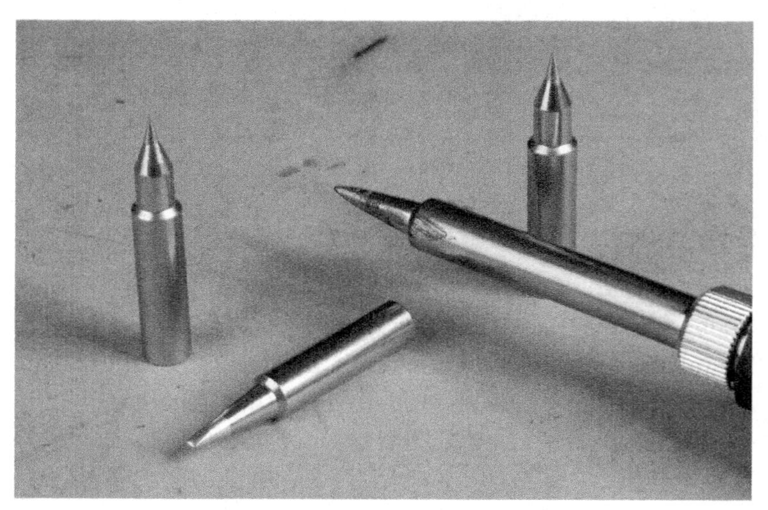

FIGURE 2-16: Hakko soldering iron tips from Adafruit Industries

There are two types of soldering sponges (shown in Figure 2-17). The most basic is simply a sponge soaked in water. The other type, which performs much better, is a brass sponge. Either one will help keep your soldering iron clean, but the brass version will not constantly cool down your soldering iron as you wipe the tip clean after soldering a joint or two.

FIGURE 2-17: Two kinds of soldering iron sponges

You will also need something to hold the sponge. A simple plate will work for a standard sponge, and other manufacturers make a nice holder for the brass sponge. If you plan on soldering a lot, go ahead and pick up a brass sponge and holder. Keep in mind that many soldering-iron holders have a sponge holder built in, as do most quality soldering stations.

> **NOTE** You might be thinking about using an old kitchen sponge as a soldering sponge, but this is not a good idea. Your kitchen sponge is not meant to be in contact with a high-temperature soldering iron. Do yourself a favor and purchase one specifically made for soldering.

Another basic item that you will inevitably need once you start soldering is a pair of diagonal cutting pliers. And, just like most tools, there is a lot of variety when it comes to the different wire-cutting pliers available. In Figure 2-18, you will see the two most common types of wire cutters: the linemen's pliers, also known as *linemen's dykes* (left); and the standard diagonal cutting pliers (right). Although both are useful in electronics, diagonal cutting pliers are much more practical for hobby electronics, since they are able to trim the leads closer to the printed circuit board after you have soldered the components in place.

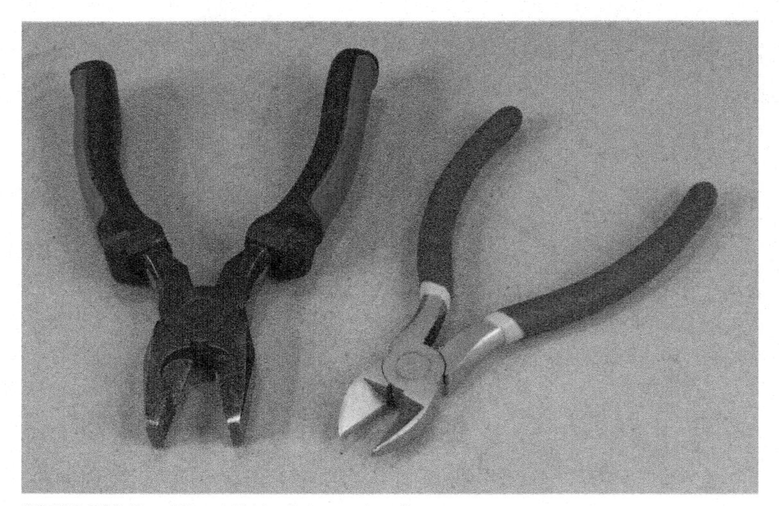

FIGURE 2-18: Two different kinds of wire-cutting pliers

If you plan on soldering more than just a few projects, or plan on soldering together a PCB, your best bet is a pair of diagonal flush cutting pliers, which are specifically made for electronics (shown in Figure 2-19). They are typically lighter and have a nice padded grip, making extended use easier and more comfortable. Also, the offset of the actual cutting head allows you to get close to the solder joint—although not too close—for trimming. They are also much better in tight spaces, which are very common in hobby electronics.

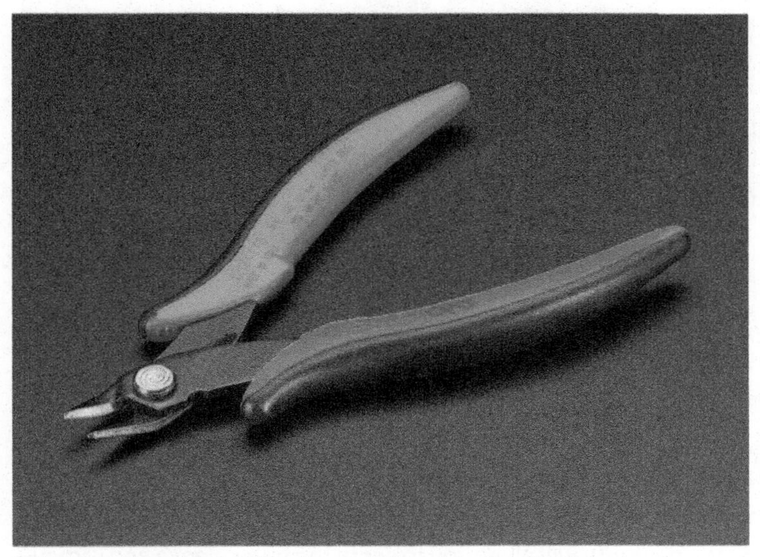

FIGURE 2-19: Diagonal flush cutting pliers are the best choice for hobby electronics. IMAGE COURTESY OF ADAFRUIT INDUSTRIES, CC BY-NC-SA 2.0.

You will also eventually need something to strip the ends of wires in preparation for soldering. There are a few options and things to consider when selecting a wire stripper. First, how much wire do you actually need to strip? If it's a lot, as in hundreds of wire ends, then your best bet is to use automatic wire strippers similar to those shown at the top of Figure 2-20. These will make short work of stripping off the exterior insulation from most types of wires. But if you don't need to prepare hundreds of wires for your next project, then the strippers in the middle of Figure 2-20 will work great. If you are on a budget, the pair on the bottom of Figure 2-20 are only a few dollars and will work just fine.

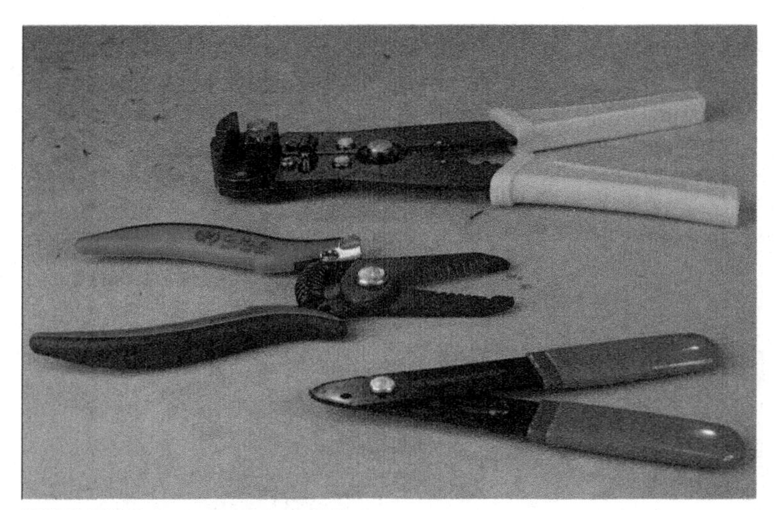

FIGURE 2-20: Various types of wire strippers

Realistically, you won't need something as robust as a pair of automatic wire strippers, like those shown at the top of Figure 2-20, when learning to solder. In fact, using your flush cutting pliers with a little dexterity will allow you to score the outer sleeve and remove the protective housing (shown in Figure 2-21). The downside to this technique is that you will, at times, cut the inner wire, possibly making it useless. This is especially true when you're stripping the wires from a component with the wire leads attached. You might have to try re-stripping the wire, but this may make it too short.

In general, when I have to strip solid core wire, I use my diagonal flush cutting pliers to make a slight cut in the housing and strip the outer layer off. However, anytime I need to strip the outer insulation off of stranded wire or any critical component, I reach for a wire-stripping tool similar to the one pictured in Figure 2-22. When you use flush cutting pliers instead of a proper wire-stripping tool, you will end up cutting some of the strands of the stranded wire, making it weaker and less able to carry current in your circuit. It's always best to strip stranded and thin gauge wires with proper wire strippers.

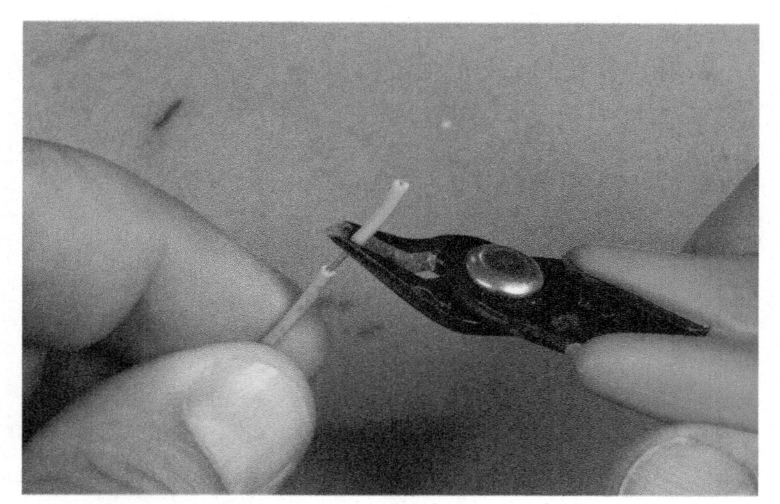

FIGURE 2-21: Using flush cutters to strip solid core wire

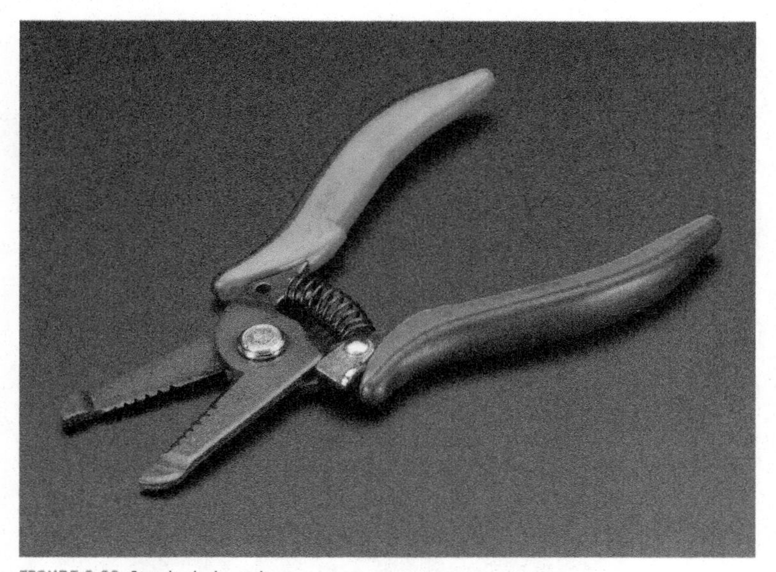

FIGURE 2-22: Standard wire strippers IMAGE COURTESY OF ADAFRUIT INDUSTRIES, CC BY-NC-SA 2.0.

> **WARNING** I've seen an untold number of people strip the ends of wires with their teeth. Don't do this! You are working with solder, which may contain lead, and it should never be ingested. Also, I have heard plenty of stories of people ending up with an emergency dentist appointment to fix a cracked or chipped front tooth!

Safety glasses are number one on my list of required soldering safety gear. Next on the list would be a proper fume extractor. If you are just getting started, sitting near a window with a well-placed fan can suffice, but it won't replace a proper commercial fume extractor (shown in Figure 2-23). A commercial fume extractor, typically consisting of a fan and a carbon filter to remove smoke, does a good job of cleaning the air. Regardless of whether you have a fume extractor, though, you should also consider the air flow of the space you are working in. A small room with the doors and windows closed is still not ideal, even if you have a fume extractor.

FIGURE 2-23: Commercial fume extractor

And, of course, just like so many tools in the Maker world, you can build your own. It's not as good at removing smoke as a commercial version, but it's portable and better than nothing. Best of all, you can practice soldering while building it! Jump to the end of this chapter to learn how to make your own mint tin fume extractor.

GETTING A GRIP ON YOUR WORK

The next tool we are going to look at is aptly called a *third hand*. It's a very handy tool, used to hold components together while soldering. It could hold two wires that need to be joined, printed circuit boards, or really anything that needs to be held securely while you make the solder connection. There are many kinds of third hands, all with a slightly different configuration. The most common has two grippers and a magnifying glass. I have never found the magnifying glass useful, and I remove it. I have other tools, which I will discuss later, that are much better for magnifying your work. The one shown in Figure 2-24 is the most basic type, and yes, I removed

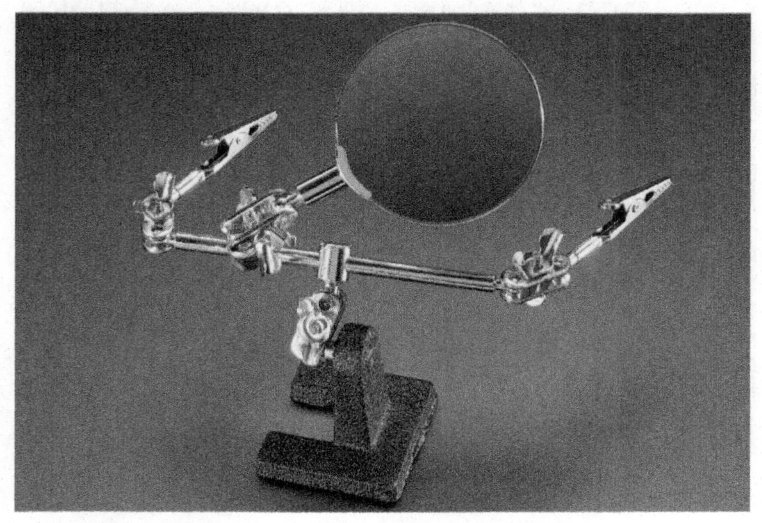

FIGURE 2-24: A basic third hand tool IMAGE COURTESY OF ADAFRUIT INDUSTRIES, CC BY-NC-SA 2.0.

the magnifying glass. Even if you decide to purchase a more deluxe version, you will still most likely want to have one of these inexpensive versions in your toolbox, as they are very "handy."

One downside of almost all third hands is the grippers. They tend to have serrated teeth that can be damaging. They may look innocuous, but they will easily cut through the outer layer of your wires, and that could lead to a short. It also makes your wires look messy! Do yourself a favor and cut a piece of heat-shrink tubing and fit it over each jaw prior to use. Figure 2-25 shows one side of my third hand with the heat-shrink tubing applied and the other without. By installing heat-shrink tubing, your third hand will have a much lighter touch on the components, which will keep it from piercing your wires. Are you not sure how to use heat-shrink tubing? I cover all you need to know about it in Chapter 3, "Let's Get Soldering."

At some point, you will most likely want a high-quality third hand to add to your toolbox. They typically have a weighted bottom, have longer arms, and just work better than the inexpensive

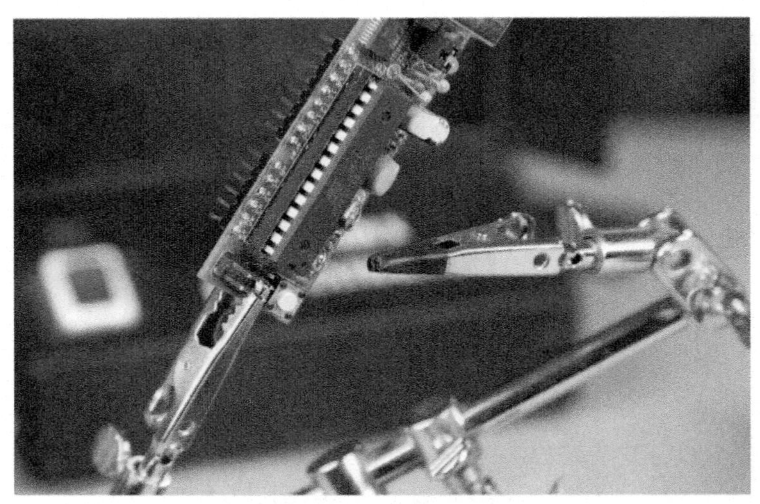

FIGURE 2-25: Adding heat-shrink tubing to the jaws of your third hand

versions. Because of these reasons, the version shown in Figure 2-26 is my go-to third hand. It has a great reach, the grippers come covered in heat-shrink tubing, and it's ready to go right out of the box. Other variations include versions with magnetic bases, more hands, and a whole host of other options. There are also tons of DIY third hand tutorials on the web. Give it a quick search and find a version that you like. Better yet, come up with your own version and share it with everyone!

When you're soldering, a third hand is very helpful when it comes to holding wires, and if needed, can be used to hold PCBs, but there is a better alternative that is well worth the investment. A PCB holder is a dedicated vise used to hold PCBs while soldering, and they are extremely helpful if you plan on soldering a lot of PCBs. The most common ones are made by PanaVise, and that's the kind I use. They make several different versions—all with varying price points. I prefer the PanaVise Jr (shown in

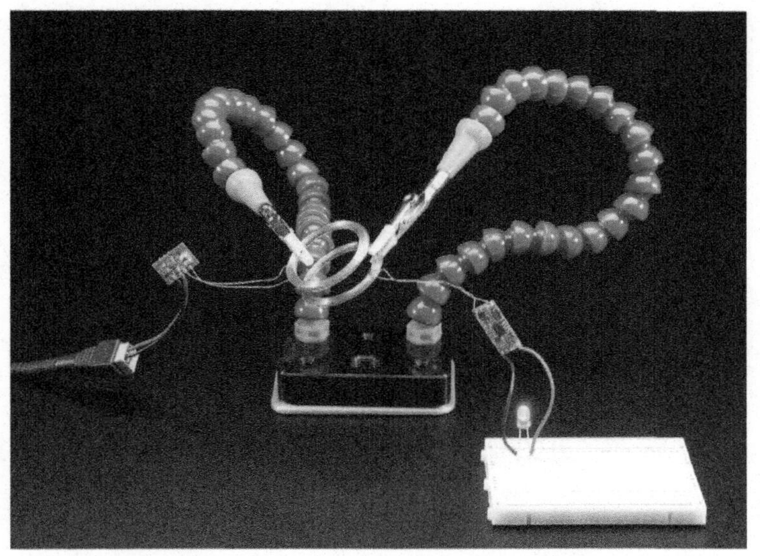

FIGURE 2-26: A high-quality third hand by Hobby Creek IMAGE COURTESY OF ADAFRUIT INDUSTRIES, CC BY-NC-SA 2.0.

Figure 2-27), and the good news is it's the least expensive. Be sure to check out all the versions of PCB holders available, then select the one that's right for you. Just like third hands, there are a lot of different types of PCB holders. Your choice will really boil down to your budget, the features you want the vise to have, such as the minimum and maximum size PCB it can hold, and various mounting options.

FIGURE 2-27: PanaVise Jr IMAGE COURTESY OF ADAFRUIT INDUSTRIES, CC BY-NC-SA 2.0.

> **TIP** If you have a PanaVise and a 3D printer (or access to one), be sure to check out Thingiverse (thingiverse.com) and search for "panavise." There are a lot of vise modifications that you can 3D print. My favorites are different handles that make the chore of adjusting the size of the jaw opening much faster. There are also files that allow you to print out an entire vise, or parts for a custom third hand tool.

A SUMMARY OF TOOLS

I tried to present many different options when it comes to picking out your equipment. It may seem a bit overwhelming, but remember that you only need a few tools to get started:

* Safety glasses
* Adjustable-temperature soldering iron with stand and sponge
* Wire cutters

Everything else falls into the nice-to-have bucket, and while it will make your experience better, it will also cost you more money upfront. It's a good strategy to get all the basics first, and then grow your electronics arsenal as needed when you tackle new projects. I picked up a few new tools just to write this book. It was a great excuse to expand my own lab! And don't forget: there are a lot of online DIY resources so you can make your own tools, like I did with my portable fume extractor. This can be a great option—you can save money and learn something, too!

MATERIALS

Next, let's look at some of the materials needed to learn how to solder. Just like in the tool section of this book, I will cover the basics. In fact, for now, *the basics* really is just talking about solder. As we learn more advanced techniques, I will cover more relevant and necessary materials like the different fluxes available and other materials needed to solder together kits, like the one pictured in Figure 2-28.

FIGURE 2-28: An electronic kit being soldered together

Solder

You might think that there can't be too many types of solder out there to choose from, and you are right—it's not that complicated. However, there are a few choices you need to be aware of when getting started (see Figure 2-29). The most difficult part when selecting a solder is to decide if you want to use a lead-based solder or lead-free. There are pluses and minuses to each, but most people I have come across use leaded solder for hobby electronics, at least in the United States. "You mean lead, as in *poisonous*?" Yes, I mean lead!

Now let me explain. Lead can lead to serious health risks, especially when ingested, so be very careful. That being said, soldering with a lead-based solder, as pictured in Figure 2-30, can be fairly safe. The main two rules are to never eat when you solder, and wash your hands immediately after you finish soldering. Your skin is a decent barrier between the inner you and the lead solder. Also, you should have good ventilation and a proper place to solder. Your dining room table is not a proper place to solder!

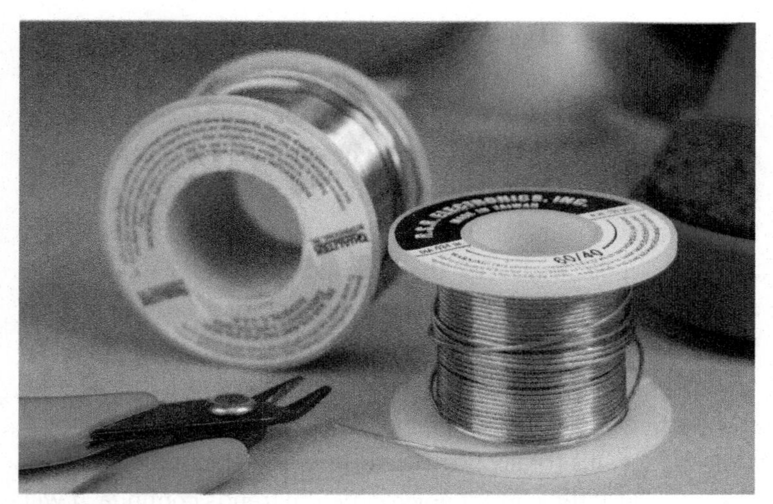

FIGURE 2-29: Various types of solder

FIGURE 2-30: A 100 g spool of standard 60/40 lead, flux-core solder IMAGE COURTESY OF ADAFRUIT INDUSTRIES, CC BY-NC-SA 2.0.

You might be asking yourself, "If there is a lead-free alternative, why aren't more people using it?" The reason is simple: lead-free, silver-based solder can be more expensive, and it is typically slightly more difficult to use, since it has a higher melting temperature. It's also prone to fatigue cracking, due to the higher temperatures needed and its inherent brittleness, and tin whiskers, which are microscopic "whiskers" that can short out your circuit. While these limitations aren't serious, I'm hoping that lead-free solder technology advances enough so there are no limitations at all.

So, which one should you use? Go with a lead-free solder to start, especially if you're teaching kids or in a public area like a Makerspace. It's totally fine for most hobby electronic projects. Additionally, if the current environmental trends continue, it will become increasingly difficult to buy or use a lead-based solder. And, as I mentioned before, there continue to be improvements in lead-free solder. So, go ahead and give it a chance. Whichever you choose, the way you make the solder connection is essentially the same.

Flux

To flux core, or not to flux core?

Once you decide between lead and lead-free solder, the next thing to consider is flux core (rosin core) or non–flux core. *Flux*, sometimes referred to as *rosin flux* or just *rosin*, is a chemical mixture that helps prevent oxidation during the soldering process. It also helps keep the solder flowing once it's melted. You will need flux when soldering, but how you introduce the flux is up to you. Using a flux core solder makes your work a bit easier, since the flux is inside the solder and will start to flow when the solder is heated. If you choose a non–flux core solder, you will have to introduce the flux by adding either a paste or liquid to each connection while soldering. Do yourself a favor and use a flux core solder when you're learning; it's just easier.

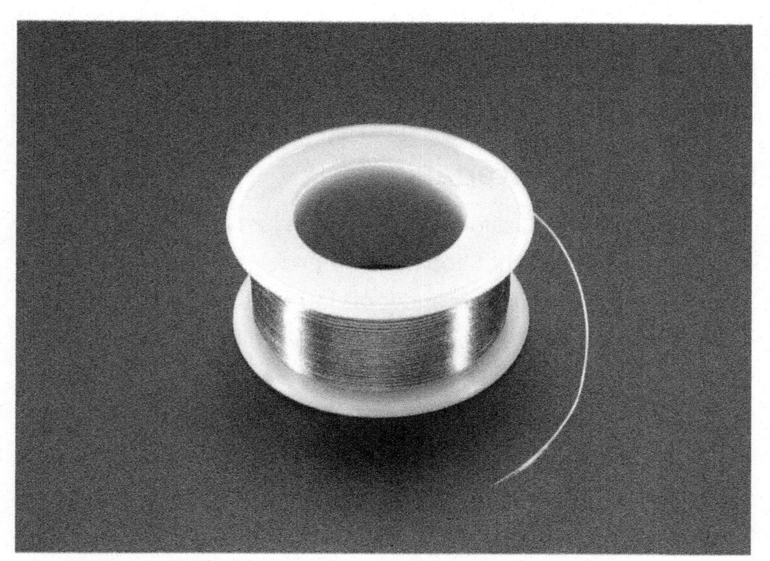

FIGURE 2-31: A 50 g spool of 0.5 mm RoHS lead-free solder

Another factor to consider when selecting solder is the ratio of raw materials used. Typically, I use a 60/40 lead-based solder. That means it is 60% tin and 40% lead, along with flux in the core of the solder. There are other ratios, which will typically vary the melting point. For 99.9% of all hobby electronics, you should choose either a lead-based flux core 60/40 solder or RoHS (lead-free; see Figure 2-31), which is sometimes labeled SAC305.

The last thing you need to consider when selecting a solder is the actual diameter, or *gauge*, of the wire. Solder is available in many different sizes, from 0.01" (0.25 mm) all the way up to 0.125" (3.175 mm) or more. If you buy a small spool, about 50 grams, of solder approximately 0.032" (0.81 mm) in diameter, it will last you a long time, and work in almost all, if not all, situations. There are times that a smaller or larger diameter would be nice, but in most cases you can do the job with anything that is about 0.032" (~0.81 mm).

Whatever type of solder you choose, don't use the kind you find in the hardware store for copper pipes! It has an acid-based flux core and it will ruin your circuits!

Summary of Materials

Just as with the discussion of tools, this is just an overview of the very basic materials that you need when learning to solder. I will cover many more materials and tools later in the book as we learn new skills and techniques (working on a project in Figure 2-32). What you really need to decide is whether or not you want to be soldering with lead-based solder. I have found that most people in the United States use lead-based solder. Different countries have different rules when it comes to the use of lead in any part of a project, so be sure to check your local rules and regulations. I'd recommend you try lead-free solder, as it certainly seems to be a healthier alternative. And no matter what type of solder you choose, make sure you wash your hands after soldering.

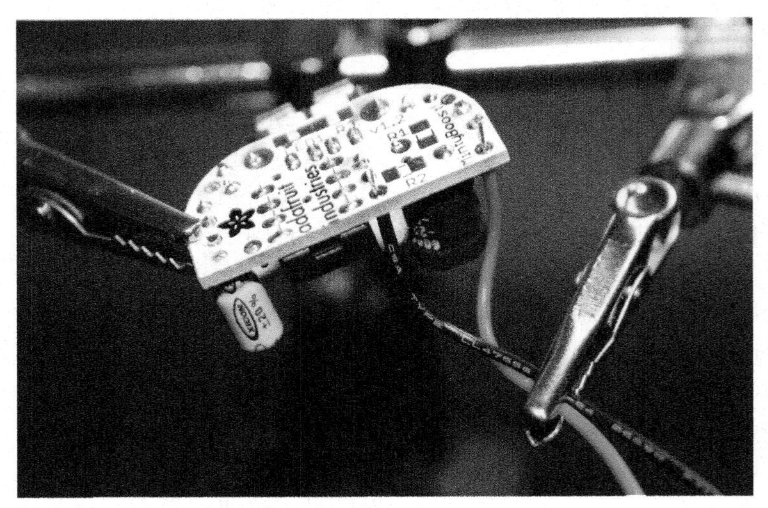

FIGURE 2-32: A kit in the middle of the soldering process

PROJECT: PORTABLE MINT TIN FUME EXTRACTOR

A fume extractor uses an activated carbon filter and fan to remove smoke and noxious fumes created from soldering. The average price of a small hobby version is about $100 USD, but the one shown in Figure 2-33 will run you about $10 USD. This fume extractor will not be as effective as a larger one, but it does a fair job and is extremely portable. Remember: always work in a well-ventilated area.

FIGURE 2-33: Mint tin fume extractor

> **NOTE** And of course, if you don't know how to solder yet, finish reading this book before you make your own fume extractor.

Tools

- Soldering iron
- Dremel with cutoff wheel
- Drill and small drill-bits
- Fine-tip marker
- Various screwdrivers
- Wire cutters
- Safety glasses

Materials

- 7812 voltage regulator
- Candy tin
- Switch
- 40 mm case fan
- 9 V batteries (2)
- "Cheap" 9 V battery connectors (2); see Steps 2 & 3
- Pieces of screen (2)
- Piece of activated carbon filter
- Some heat-shrink tubing
- A few inches of wire
- Flux core solder
- Miscellaneous screws and washers
- Paint (optional)

> **WARNING** Always wear safety glasses when soldering, when drilling, and especially when cutting metal!

Step 1: Build the Circuit

I decided that making a quick mock-up might be a good idea, and I am glad I did. At first, I thought that running the case fan off of just one 9 V battery would provide adequate power. In the end, though, I decided that 12 V "sucked" better—and in this case, that's a good thing.

The final circuit, shown in Figure 2-34, uses a simple switch, two 9 V batteries, a 40 mm case fan, and a 7812 voltage regulator. The 7812 takes the 18 V from the two 9 V batteries that are wired in series down to 12 V, which is what the fan requires.

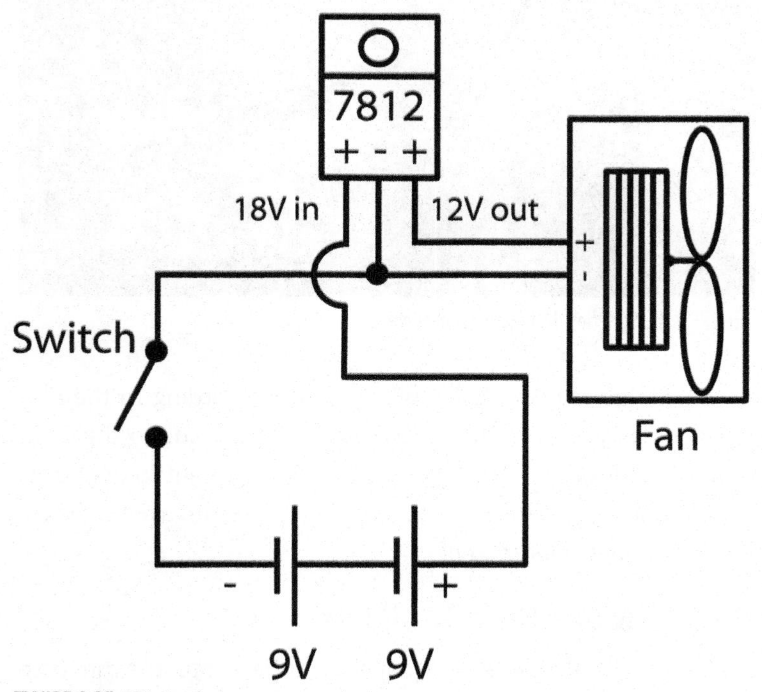

FIGURE 2-34: DIY mint tin fume extractor circuit

Step 2: Solder the Components

Notice the battery connectors in Figure 2-35, which are the cheap, flexible vinyl version, not the hard-plastic type. This allows them to easily fit in the case. The difference in thickness is minimal, but it is enough to stop you from putting both 9 V batteries in the case.

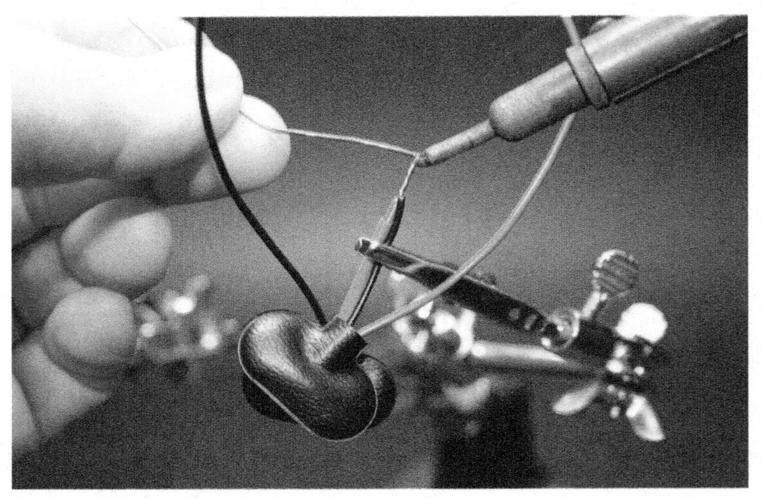

FIGURE 2-35: Soldering the battery connections

This is a very simple circuit. Solder it according to the diagram, making sure to attach the component leads to the 7812 in the proper order. (See Figure 2-36.) Don't forget to use heat-shrink tubing on all the connections. This goes into a metal box—metal conducts electricity!

Step 3: Make Sure It all Fits

Make sure that everything can be stuffed into the tin (see Figure 2-37).

FIGURE 2-36: The completed circuit all soldered together

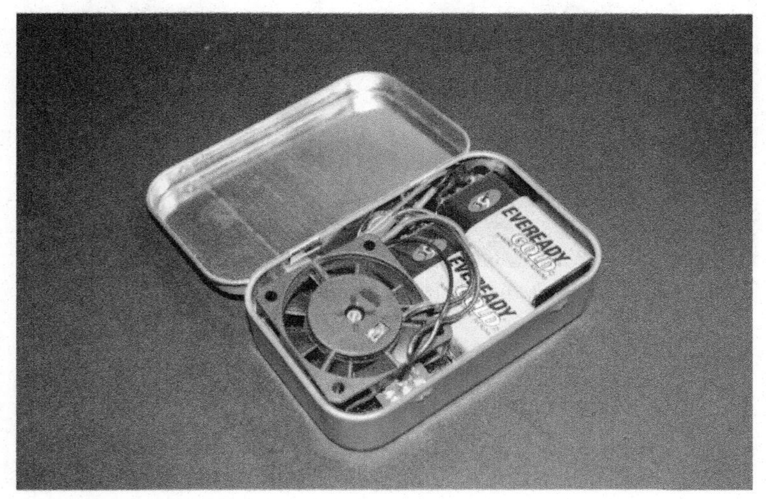

FIGURE 2-37: Making sure it all fits in the tin

Step 4: Cut and Mark the Openings (Wear Safety Glasses!)

I used a marker and paper to make a template for the fan openings, which are 35 mm square. I marked the opening for the switch at the same time. Then, I cut all the openings with a Dremel tool and cutoff wheel (wear safety glasses!). Next, I marked and drilled the two holes for the switch screws and the hole for the regulator (see Figure 2-38).

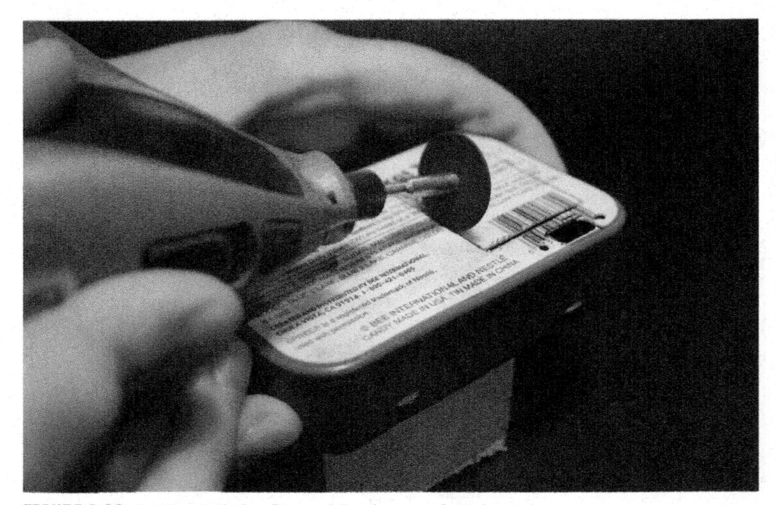

FIGURE 2-38: Cutting the holes. Be careful and wear safety glasses!

After you cut the first fan hole, close the box and use the 35mm square paper template to align the second hole. Make sure to allow for air flow, but you can just eyeball the placement (see Figure 2-39). In reality, it isn't that crucial for them to line up exactly. There *is* room for error.

Step 5: Paint

I decided to paint the tin with a nice, red spray paint (shown in Figure 2-40). I hot-glued a scrap piece of wood to the inside so I could hold it while I spraypainted it. I gave it two quick coats, and I think it looks good. Krylon paint is fairly toxic and flammable, so paint it outside and away from everything!

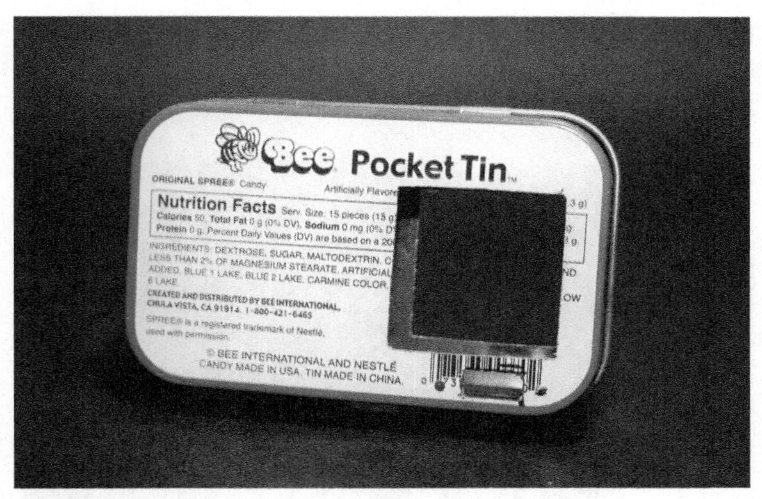

FIGURE 2-39: Make sure the holes line up for proper air flow.

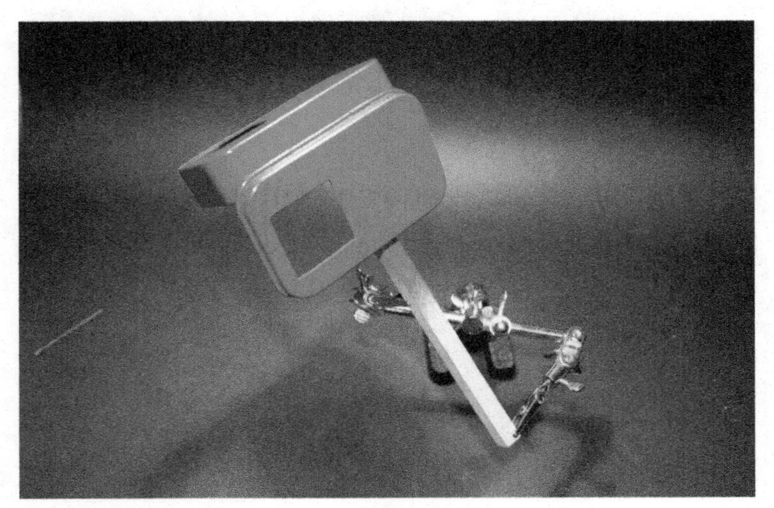

FIGURE 2-40: A nice coat of paint is well worth the effort.

Step 6: Attach the Regulator and Switch

First, screw in the 7812 using a screw and a washer or two to add some space between it and the side of the tin. I used a #6-32 screw and one washer to keep it away from the side, but you can use anything that fits. The screws and washer will also act as a heat sink. Finally, screw in the switch. (See Figure 2-41.)

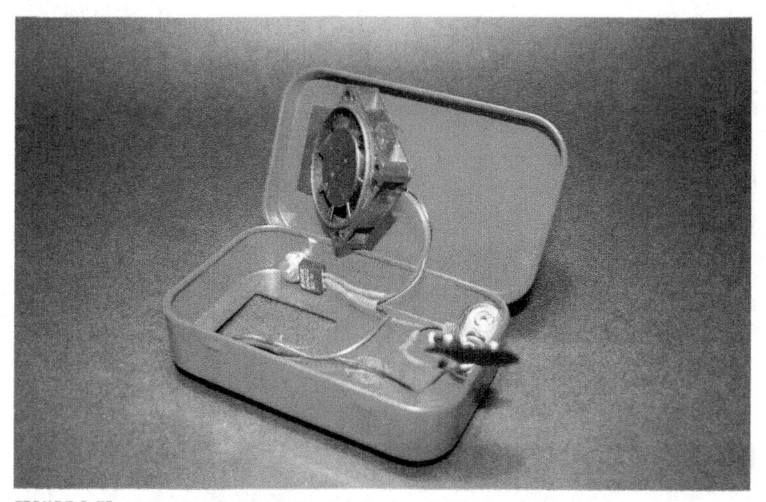

FIGURE 2-41: All the electrical components being installed

Step 7: Add the Screens and Filter

Here you can see the screen-filter-fan-screen sandwich (Figure 2-42). The screens are 50 mm^2 and the filter is 40 mm^2. You can buy replacement filters for the commercial extractors at a reasonable price.

FIGURE 2-42: Filter, fan, and screen sandwich

Next, just hot-glue or epoxy the corners of the screens down, and sandwich the filter and fan in between. Compression will ultimately hold it all together. (See Figure 2-43.)

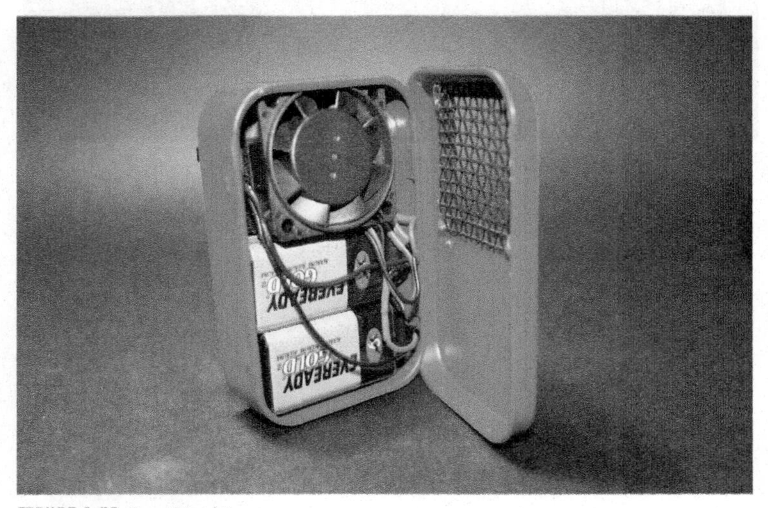

FIGURE 2-43: Everything fit!

Step 8: Test

I have run mine continuously for hours and have had no heat buildup from the 7812, and the fan is still running strong. (See Figure 2-44.) It seems to work quite well and, although it is no replacement for a larger commercial fume extractor, it will come in handy for small projects. Remember: follow all safety guidelines when soldering, and work in a well-ventilated room, even if you have a fume extractor.

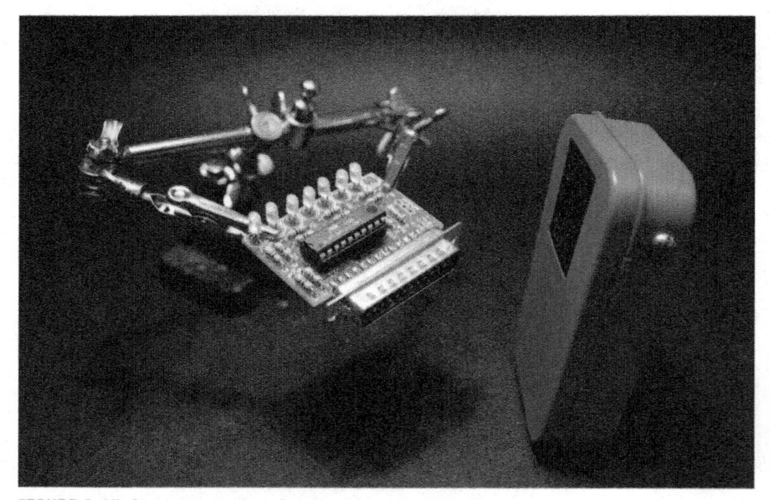

FIGURE 2-44: Give it a try and see how it performs.

3

Let's Get Soldering

Finally, we're at the part where you get to fire up your iron and learn to solder. It might seem like we went through a lot to get to this point, and you're right—we did. Just remember that all the information, tips, tricks, and recommendations came from years of experience. My hope is that I will save you a lot of time and money and, more importantly, make the learning process fun by helping you avoid some of the bad tools and materials that I have seen over the years.

Simply learning to solder takes just a few minutes, especially when you have the right tools. Thousands of people have learned to solder at Maker Faires all over the world, and it took less than 10 minutes for most of them to learn the basics. OK, let's get started!

Stay Safe!

Soldering *can* be dangerous. Never attempt to solder anything without a thorough knowledge of the process and all the possible hazards involved. Here are a few key points to remember when soldering:

- **Never solder a live circuit**, any circuit that is connected to a power source, or a circuit that has residual power in it (for example, a circuit with a capacitor that is still charged).
- **Never solder anything connected to mains electricity**, also known as *household electricity*. This should only be done by professionals because the high voltage and current are deadly. **If it plugs into a wall, don't mess with it!** Ever!
- Lead solder is toxic. **Always wash your hands after using any solder**, especially lead-based solder, and avoid breathing in the fumes. Never use lead-based solder where it will come in contact with food or drink or leave it where children might play with it.
- **Always wear safety glasses** when soldering.
- **Never solder anything directly onto a battery!** It can explode!
- Always have **proper ventilation** when soldering.
- Always ensure you're in a **proper location** when soldering.
- Keep a **fire extinguisher** readily accessible when soldering.
- The techniques covered in this book are for **hobby use only**. Any critical soldering, or soldering in unfamiliar situations, should always be performed by a professional. If you are unsure of anything, ask a professional!

PREPARING YOUR SOLDERING IRON AND WORKSTATION

When you start your project, it is very important to have a safe place to solder with good light and adequate ventilation. Remember: your kitchen table isn't a proper place to solder. However, if you need to solder somewhere that is less than ideal, you can always put down a scrap piece of plywood to protect the surface you are working on. If you are using lead-based solder, do not work anywhere with food preparation in its future. You can see in Figure 3-1 that I have all the required tools and materials ready to go. You don't want to be running around looking for things once your iron is hot—you never want to leave a hot soldering iron unattended.

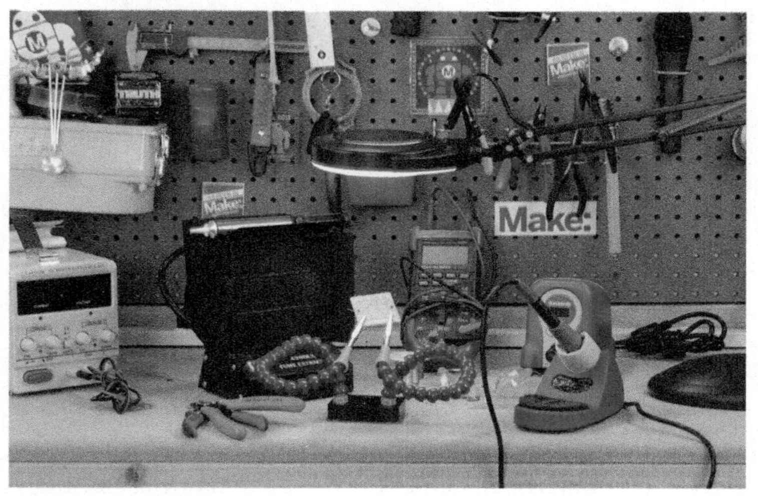

FIGURE 3-1: A proper spot to solder

Before you start using a brand-new iron, you will need to prep the tip, which is called *tinning the tip*. (See Figure 3-2.) This allows the solder to flow more easily when making connections, and it protects the soldering iron tip from corrosion. It's a simple process. Just turn your iron on to about 650°F (340°C) if you are using lead-based solder, or 700°F (370°C) if using lead-free solder. Once the iron is hot, simply melt some solder on the tip, making sure it's completely covered. Once it's covered in solder, give it a quick wipe on your solder sponge or brass sponge. It should look nice and shiny, and be ready to use.

FIGURE 3-2: A new soldering iron tip waiting to be tinned

This process of cleaning and tinning the tip should be repeated as much as necessary to make sure your soldering iron tip is always clean. I wipe mine about every three solder connections, and every time I pick up my iron or put it back on the stand. If you plan on not soldering for more than 10–15 minutes, turn your iron off. This will help minimize the oxidation that will inevitably occur on the tip of your iron. One of the biggest mistakes I see when someone is learning to solder is attempting to solder with a very dirty tip (shown in Figure 3-3). It just makes for a bad connection and experience. One last thing to remember is that after you are done soldering, go ahead and tin and wipe the tip again (see Figure 3-4). This will leave a nice protective coating of flux and solder on your tip while not in use.

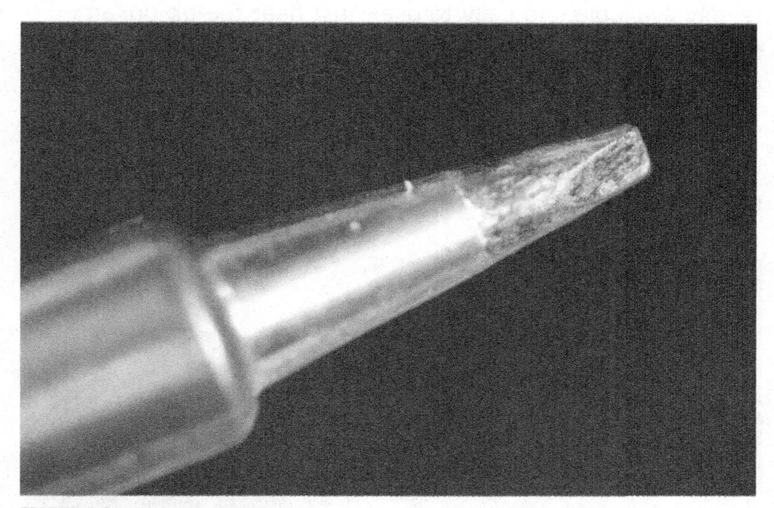

FIGURE 3-3: A dirty soldering iron

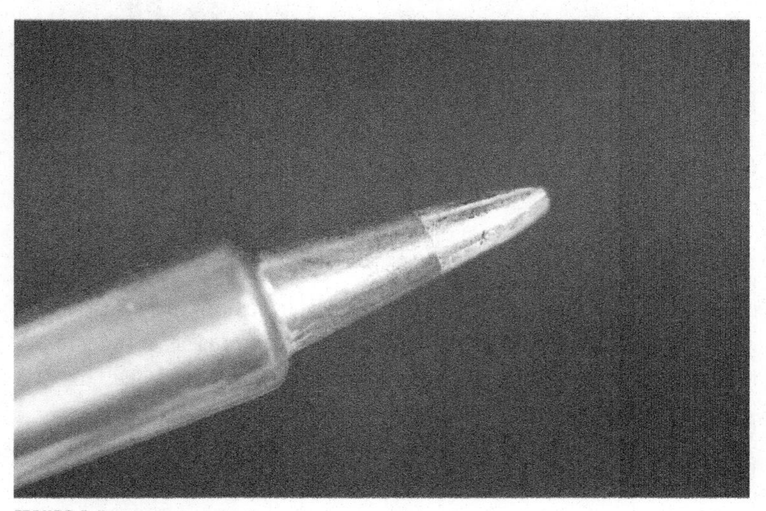

FIGURE 3-4: The tip after cleaning

If your soldering iron tip gets really corroded, you can use tip-tinning paste to try and restore the tip (see Figure 3-5). It's readily available and easy to use—just heat the tip up, and dip the tip in the tin repeatedly. It should restore the tip with just a few applications. If not, you'll need to install a new tip on your soldering iron. It works so well that I also like to use it once in a while just for regular maintenance. A small, inexpensive tin will last for years.

FIGURE 3-5: Using tip-tinning paste

> **WARNING** Never use a file or sandpaper on the tip of your soldering iron. It may be tempting if the tip has excessive corrosion, or to try to make the tip smaller for different applications, but it's a bad idea. Don't ever do it! You will remove the outer coating of the tip and it will never conduct the heat as it should. If your soldering iron tip is so corroded that tinning the tip won't fix it, go ahead and purchase a new one.

SOLDERING PRINTED CIRCUIT BOARDS (PCBS)

It seems like most people dive right into soldering printed circuit boards. I think it's because there are so many amazing kits to solder together—from simple blinky LEDs to Internet-connected devices, and more. There are thousands of different kits made by hundreds of different kit makers, like myself, so it's no wonder so many Makers want to learn to solder. Regardless of the reason, it's a popular first soldering project, so that's where I'll start.

If you are soldering a few wires together, or the components you are using aren't sensitive to static, you may not need an antistatic wristband (shown in Figure 3-6). It's always a best practice to ground yourself to avoid any static that may be present, but the truth is that most people don't use them in hobby electronics—I have one, but rarely use it. If you are working in a lab or a professional environment, you will see them in use everywhere. They are extremely important when working on expensive integrated circuits (ICs) or complex or expensive boards, or when there is

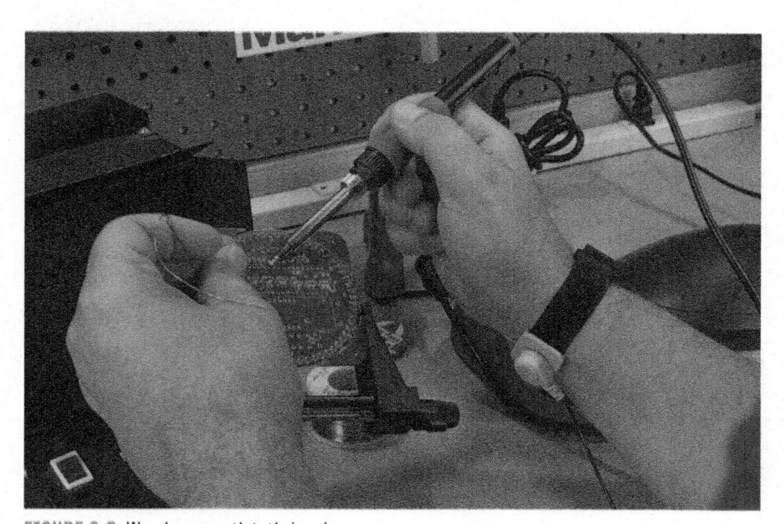

FIGURE 3-6: Wearing an antistatic band

a threat of static electricity. I'm willing to take the chance of not wearing one when soldering a project, mostly because I would use a socket that allows me to plug in the IC after all soldering is done. This also allows me to replace the ICs in the future, as needed, without having to desolder the IC.

In most cases, your PCB will be ready to use out of the box, but I have seen some that have enough corrosion from manufacturing or extended storage that they needed to be cleaned. Fortunately, dirty boards are not too common, and if you get one, they're easy to clean. I keep a few electronic-specific isopropyl wipes handy for just this purpose. Just wipe the board down, as shown in Figure 3-7, and be sure to properly dispose of the wipe. There are also dosing bottles and pads that you can pick up for this purpose, but those are usually only found in a facility that is soldering a lot of PCBs.

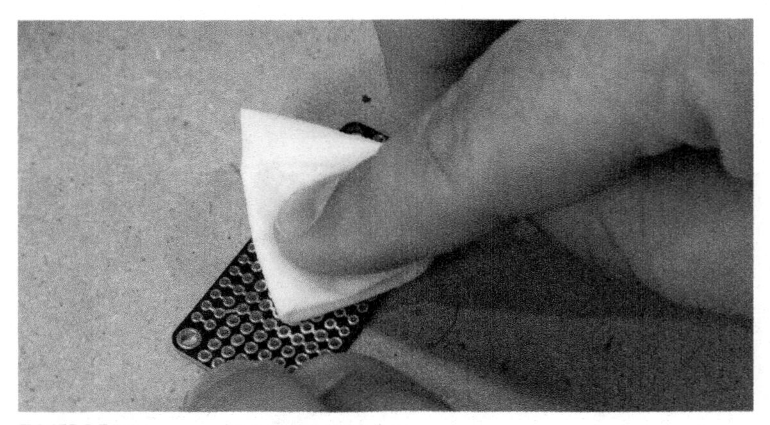

FIGURE 3-7: Wiping your soldering connection clean

> **WARNING** Isopropyl alcohol wipes are extremely flammable. Be sure to read all the warnings from the manufacturer prior to use. Also, do not use them near a hot soldering iron, or any source of heat.

First, you will need to place the parts on the PCB. We're not going to cover the specifics of how to do this, because the instructions that came with your kit should cover those details. Instead we'll start with how to make the mechanical and electrical connections once the parts are placed. A common mistake people make when starting out is not properly holding the parts in place while they are preparing to solder. You should insert the part into the board and bend the wires at about a 45° angle, as shown in Figure 3-8, which will keep the part in place while you solder the connection.

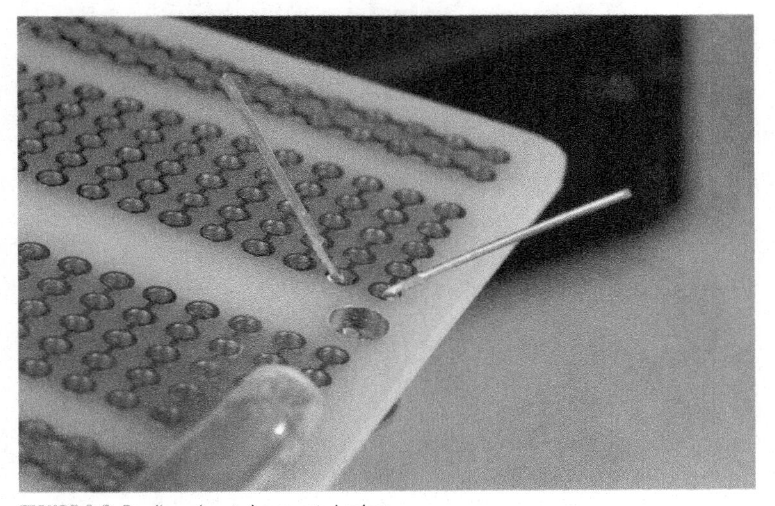

FIGURE 3-8: Bending wires to keep parts in place

> **NOTE** Make sure the component is flush with the front of the PCB. Sometimes the part will slip and end up too far away from the PCB. If this happens, you can typically push it back into place with some tweezers or pliers while heating up the solder.

Most beginners tend to bend the wires flat to the back of the PCB, as shown in Figure 3-9. Do not do this! This will cause electrical shorts, and make the soldering and trimming of the leads much more difficult. You only need to bend them enough so that the part doesn't move when you solder the connection. Remember: in most cases, you will flip the board over to solder the component in place. If the wire leads aren't bent, the part will fall out. Skip to Chapter 4, "Troubleshooting and Fixing Mistakes," to read more about how to fix mistakes like this one.

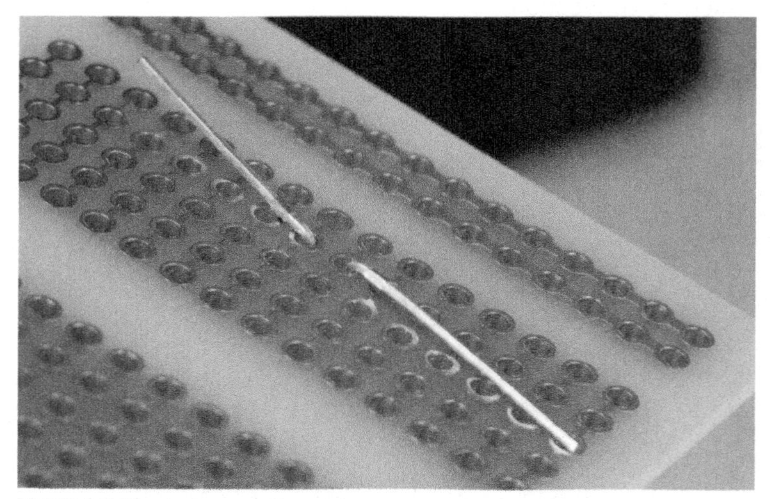

FIGURE 3-9: Wires bent too far down, which will lead to short circuits–don't do this!

If your component, such as an IC socket, doesn't have long leads, you can use tape to temporarily hold it in place. Tape will work if you don't have any other means of holding a component, but make sure it's attached to the PCB far away from where you are soldering so it doesn't melt. Melted tape will leave behind residue that will need to be removed prior to soldering any more components. A much better option is to use copper clamps or alligator clips, as shown in Figure 3-10. You can find these clips in various sizes and made from various materials. It's a good idea to have a few in your lab for tricky soldering jobs.

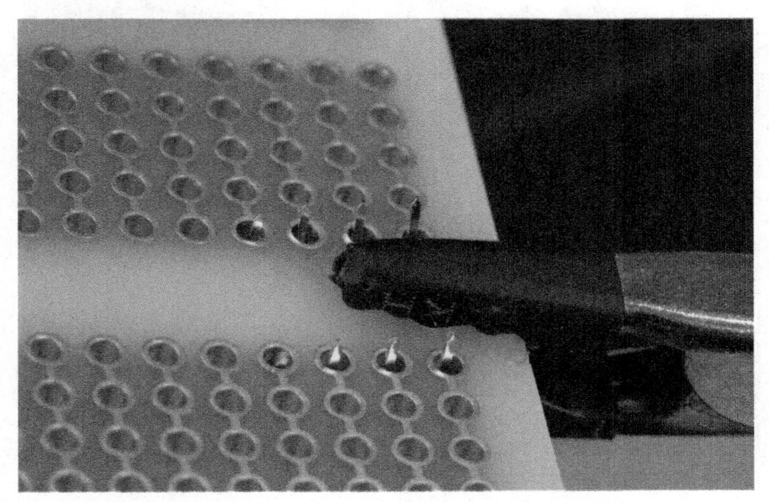

FIGURE 3-10: An alligator clip in the process of being used to hold a component in place

Now it's time to solder! Make sure your iron is up to temperature. If you are unsure, you can touch the tip of the iron briefly to the solder to see if it melts. It should melt almost instantly. Be sure to wipe the tip clean prior to soldering.

The key to a good solder joint is the placement of the tip of the iron (shown in Figure 3-11). It should be at the intersection of the PCB and the component lead. You should heat both parts of the joint so the solder flows to both the component and the PCB equally. How long you keep the iron in the joint varies a little bit based on the temperature of your iron, but no more than a second or two. In general, you can heat the connection for two seconds, then feed a little solder to the joint for about one second—again right at the intersection of the component and PCB—then remove the iron and allow the solder to cool.

FIGURE 3-11: Proper placement of the soldering iron tip and the location of where to feed the solder

> **NOTE** I usually tell people to think of this little process when making their first connections:
>
> 1. Apply the soldering iron, then count "one Mississippi, two Mississippi."
> 2. Now, feed the solder into the joint, count "three Mississippi, four," and remove the iron.

The entire process should only take about three to four seconds total. Any longer, and you might burn the PCB or the component, and the flux will start to lose its integrity and no longer work.

NOTE Choosing a temperature for your iron is not an exact science. I like to solder with a hot iron, which makes the solder connections faster. When starting out, though, you should use a lower temperature. Typically, when using a 60/40 lead-based solder, you should set your iron to 650°F (approximately 340°C) or a little higher for lead-free solder—700°F (approximately 370°C). The hotter the temperature, the quicker the solder will start to flow. As you become a more proficient at soldering, you will tend to use a hotter iron.

You will know that your timing is right when you see the "cone of perfection." The solder joint should look fairly shiny and have the appearance of a little volcano, as shown in Figure 3-12. When solder is heated with flux to the ideal temperature and becomes fluid, you are creating a property of wetting at the connection point. *Wetting* is what happens when the solder penetrates the copper trace of the PCB and creates a new alloy. Proper wetting of the solder will not only create an ideal electrical connection between your component and PCB, but also create the strongest mechanical connection possible.

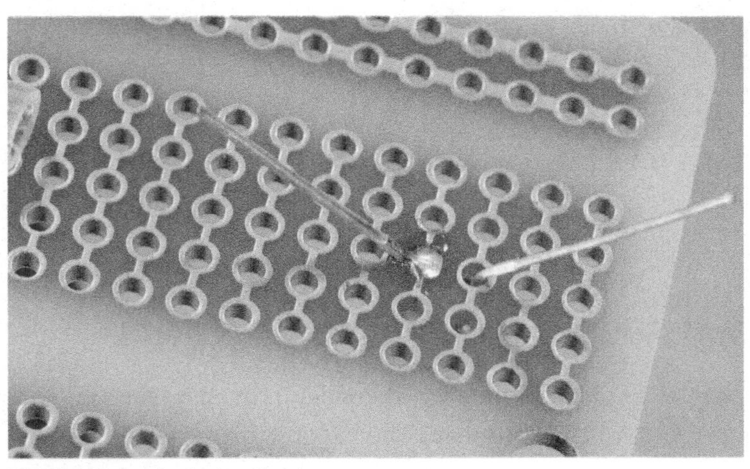

FIGURE 3-12: A perfect basic solder joint

If your solder joint doesn't look like Figure 3-12, then skip on over to Chapter 4, where you will see examples of what might have gone wrong and how to fix a bad solder connection.

Some components, like USB ports and power jacks, have large lugs to help make an extra strong mechanical connection between them and the PCB (see Figure 3-13). Be sure to always add enough solder to these components, because they will experience stress every time they are used. You'll be surprised how much solder can flow into these connections. Be ready for this by unspooling enough solder to fill in the holes of the PCB before you start.

FIGURE 3-13: A USB connector with four leads and two lugs

TRIMMING THE WIRE LEADS AFTER A SUCCESSFUL SOLDERING CONNECTION

Once the solder connection is made, in most cases you will need to trim the wire lead. The exception to this rule is any component where the lead is short enough to not interfere with other components or the housing of your project. On the left of Figure 3-14, you can see the properly trimmed leads of an LED, and on the right you can see two rows of leads from an IC that didn't need to be trimmed. When you're trimming the wire leads, do not cut them flush with the PCB. You should trim the leads at the top of the solder cone, or about one-third of the way down the solder cone—no more. You do not want to cut them too close, as this will weaken the joint and possibly cause it to fail in the future.

FIGURE 3-14: Wire leads untrimmed (left) and an IC that doesn't need trimming (right)

On the left of Figure 3-15, you can see a solder connection that was trimmed way too close to the PCB. This may weaken the electrical and mechanical connection in the future. There is no need to trim the leads this close. On the right, you can see a lead that could use a bit more trimming. It protrudes enough that it could get hung up on another component or interfere with installing it in the plastic housing.

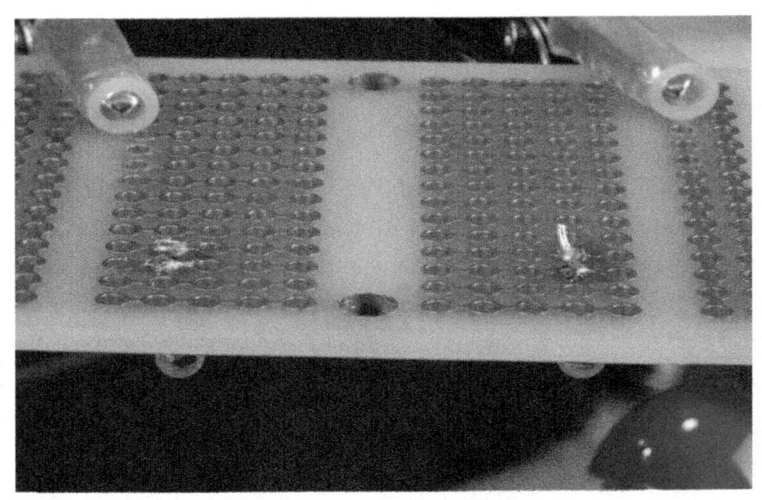

FIGURE 3-15: Too close of a trim (left) and too much wire lead (right)

CONNECTING WIRES OR COMPONENTS WITHOUT A PCB

Not all soldering occurs at the intersection of a PCB and a component lead. Many times, you will need to solder some wires or components directly to each other. Although this is a simple process, I will share some tips and tricks to make it a perfect connection.

It can be very tempting to solder wires together by laying them directly on your work surface. Don't do this! You will almost certainly burn your work surface, make the tip of your iron dirty, and worst of all create a disturbed solder joint that is mechanically unsound. Instead, use your third hand to hold the wires securely in place.

Prior to soldering, make sure to twist the wires together, making a solid connection. Then, flood the connection with plenty of solder.

In Figure 3-16 you can see the heat-shrink tubing on the jaws of the third hand. This prevents the teeth from slicing into the exterior insulation of the wire, which might cause a short or allow moisture into the connection. If the wires are attached to other components, don't forget to add the heat-shrink tubing to the wire prior to soldering the connection. (More about heat-shrink tubing a little later on.)

FIGURE 3-16: Wires ready for soldering

NASA has high standards when it comes to splicing two wires together. They use a process called a lineman's splice, and although it's unlikely that you are building something that requires such scrutiny, when it comes to soldering a connection between two wires, it's tough to beat the lineman's splice, shown in Figure 3-17. In the figure, you can see the pre-tinned wires pass each other, and then wrap three times around each other, forming an incredibly robust connection. (You should always wrap the wires a minimum of three times.)

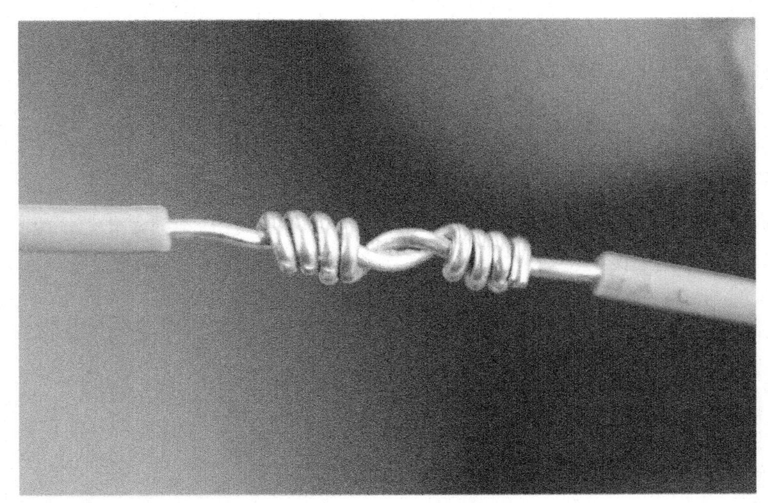

FIGURE 3-17: NASA lineman's splice

Once you wrap the wires, trim the ends flush and cover the entire area with a generous amount of solder, as seen in Figure 3-18, then cover the connection with heat-shrink tubing, as in Figure 3-19. If you want your wire connections to last a long time, take some advice from NASA and use a lineman's splice.

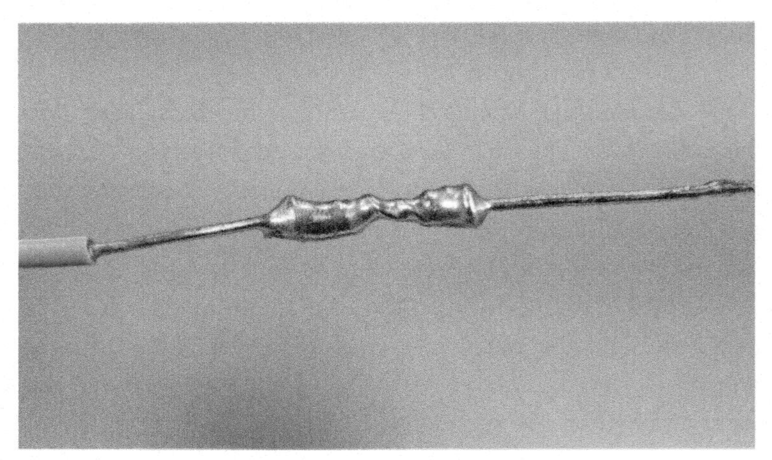

FIGURE 3-18: Wires after being soldered together

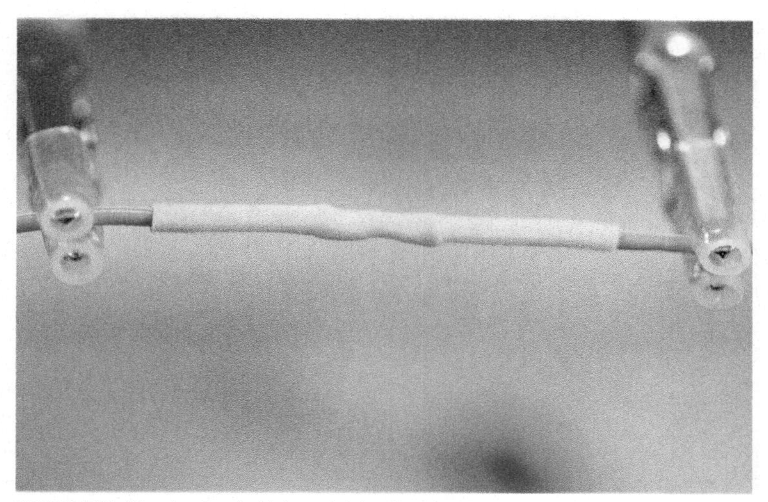

FIGURE 3-19: Wires insulated with heat-shrink tubing

NOTE If you are relying on the lineman's splice shown in Figure 3-17 to build a rocket, it will not go well for you. There are a lot of details that go into building critical equipment for NASA. The instructions are intended to teach you how to make a more robust connection, not to build a rocket! If you want to learn more about the fascinating requirements of building projects for NASA, check out *https://go.nasa.gov/2hlhHw3* or search the Internet for "NASA workmanship standards."

Some components, such as motors and switches, will have terminals that you will need to make connections to (as shown in Figure 3-20). In general, your best bet is to use stranded wire, since it's more flexible and will perform better—but solid-core wire will work, too. Do not just place the wire on the terminal and solder, which is something I see far too often.

FIGURE 3-20: Soldering wires to the terminals of components

Take the time to fish the wire through the terminal and do the lineman's splice, which we learned how to do earlier and is shown in Figure 3-17, and applied to a switch, as shown in Figure 3-20. This will give you a lot of surface area to make a good mechanical connection. If there is no hole for the wires to go through, you can pre-tin the terminal with some solder, then place the wire on the terminal and heat the connection after wrapping it. In either case, the solder will flow onto the wire and terminal, making the connection (see Figure 3-21).

> **TIP** Sometimes, I see people flood the wire with solder, but they don't heat the terminal enough, preventing the wetting action at the connection. This will result in a loose electrical and mechanical connection that will most likely fail. Make sure the solder flows into the wire and over the terminal of the component, and heat the wire and terminal together so the solder will flow evenly.

FIGURE 3-21: Wire successfully soldered to the terminal of a switch

FIGURE 3-22: Wire insulated after soldering

Once you make a connection, especially when it's a splice between two wires, you need to protect that connection from shorts and corrosion. You can use electrical tape to secure it, but that typically fails, and it can add a lot of bulk, which can be undesirable for hobby electronics where space is limited. Your best option for sealing the connection is to use heat-shrink tubing (shown in Figure 3-22). There are several different kinds of heat-shrink tubing, but most function the same—when heat is applied, the tubing will shrink to approximately half its diameter.

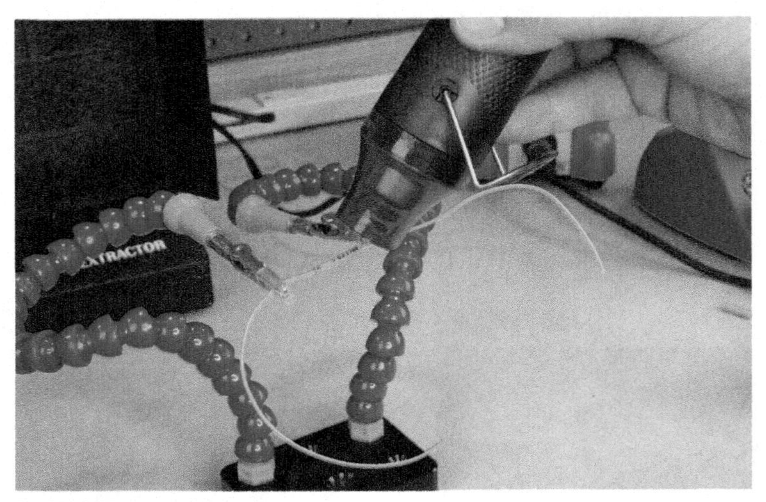

FIGURE 3-23: Heat-shrink tubing

> **TIP** Don't forget to slip the heat-shrink tubing onto the wire splice prior to soldering. It's never fun to have to cut a fresh connection because you forgot to add heat-shrink tubing!

Start by selecting a piece of heat-shrink tubing that is longer than the exposed wire, and slightly larger than the connection. You will need to slip it onto the wire and away from the connection before soldering. After you have soldered the wires, move the heat-shrink tubing over the bare metal of the wires and apply heat to it. This will make it shrink, and prevent shorts in your circuit. It's fairly common to use a lighter to heat the tubing, but I encourage you to use a hot-air tool instead. I've seen tubing catch fire and burn far too often. The type of hot-air gun I'm using in Figure 3-23 costs less than $20 USD and it's a lot safer than an open flame.

> **TIP** Make sure to trim the leads of your wires prior to soldering, as they can poke through the heat-shrink tubing, causing shorts.

Another alternative would be to use liquid electrical tape. Although this works, I find that it makes a bit of a mess and don't use it too often. If I'm going to be totally honest, I usually resort to using it after I have made a connection and forgot to put heat-shrink tubing on the wires first. Don't laugh too hard; it will happen to you too!

4

Troubleshooting and Fixing Mistakes

The process of soldering is fairly simple, but a few things can and will go wrong. Fortunately, most of these issues are avoided or easily fixed. After some practice, you will be able to solder hundreds of connections without any issues whatsoever.

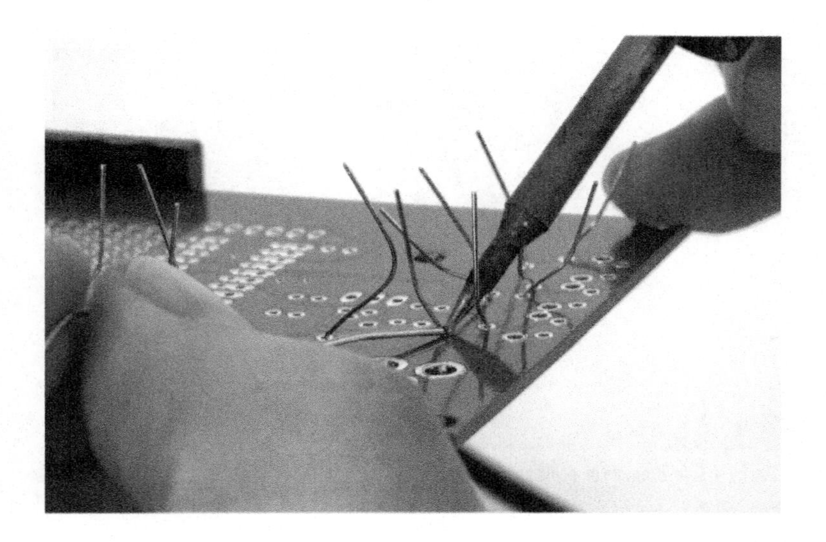

COMMON MISTAKES

In this section, we'll look at some of the most common mistakes that are made by people learning to solder, how to recognize them, and solutions for fixing them. Even experts sometimes make these mistakes, so don't feel bad if you do, too. Everything gets better with practice!

Cold Joint

The *most* common mistake that I have seen people make is not heating components equally, or enough, which creates a *cold joint*. This happens when the component lead is not entirely heated by the iron, or the iron is on the lead for too short a time or at too low a temperature, as shown in Figure 4-1. This makes the solder bond to the lead, but not to the joint of the PCB. The same can happen if you heat the copper pad of the PCB more than the lead wire, which results in the pad receiving enough solder, but not the lead. In either case, the wetting action, which forms a proper electrical and mechanical connection, did not occur. You can typically fix this mistake by reheating the joint closer to the side where the solder did not flow and trying to make the connection

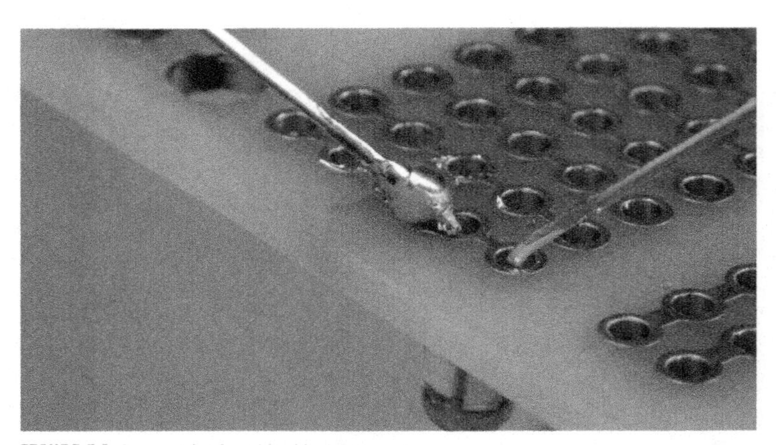

FIGURE 4-1: An example of a cold solder joint

again. You may need to introduce a little more solder if the joint looks dirty from oxidation.

> **TIP** If the connection you made is oxidized (looks "dirty"), which may be caused by overheating or several other reasons, you can add some flux or a little additional flux-core solder to help the solder reflow evenly. You can purchase flux pens or paste for just this reason. When you need to reflow a joint, a flux pen is a lifesaver. Just dab a little flux on the joint and reflow the solder—simple.

Too Little Solder

Another very common issue is not flowing enough solder into the connection (see Figure 4-2). I see this happen to people when they are nervous about soldering. They think the board will burn if they heat it for too long, so they just feed a little solder in and remove the heat. Although it is true that you can burn the board if too much heat is applied for too long, not adding enough solder makes for a bad mechanical connection. To fix this mistake, simply reheat the joint and introduce some more solder.

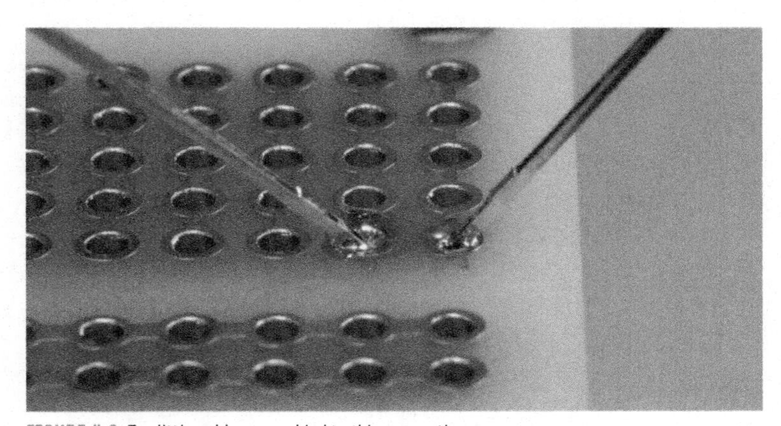

FIGURE 4-2: Too little solder was added to this connection.

Too Much Solder

Too much solder can also be a problem (see Figure 4-3). Although you might get away with this common mistake if it's only at the joint, it can cause issues. For example, if the solder touches another lead or pad of the PCB, it can short out the leads of the other component. If you flow too much solder in a connection, it's easily removed with either copper braid or a solder sucker, a technique that is covered later in this chapter.

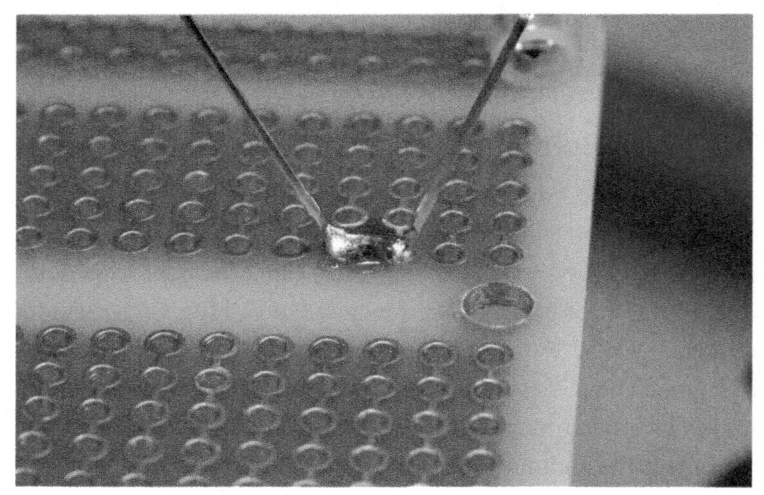

FIGURE 4-3: Too much solder causing a solder bridge

> **NOTE** Removing solder is a solution to several common problems you may encounter when learning to solder. You can learn about a few different ways to do this by skipping ahead to the section "Different Ways to Remove Solder," later in this chapter. This is an important skill to learn, because you *will* make mistakes. Fortunately, removing solder isn't too difficult if you use the proper technique.

Too Much Heat

Although it's more common to apply too little heat at the solder joint, too much heat can be just as much of an issue (see Figure 4-4). It can have a negative effect on the integrity of the connection, and it can also melt or burn the PCB, both of which are not good! In some situations, the copper pad of the PCB can be destroyed beyond repair, although this is rare. Lastly, too much heat can damage the component being soldered, especially if it's an integrated circuit (IC) or other heat-sensitive component. Too much heat can be difficult to fix, so prevention is the best course of action—use less heat. It's much easier to fix a cold joint by reflowing the solder and making a perfect connection.

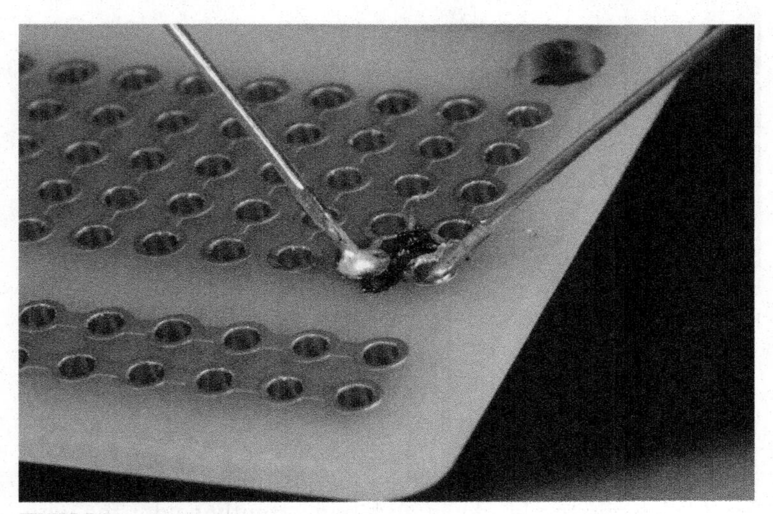

FIGURE 4-4: Too much heat at the connection

Too Much Movement

You need to make sure the parts you are working on are held securely, and this is why you need a third hand or PCB vise. If the components move while the solder is hot, you will create a joint that looks rough, as seen on the left of Figure 4-5. This is commonly referred to as a *disturbed joint*. This may not make a reliable electrical connection, and it can also make for a more brittle joint, which could lead to future mechanical failure. You can usually fix this type of poor connection by reflowing the solder at the joint, while making sure nothing moves. Again, you may need to introduce a little extra flux or solder to make sure the solder flows and a proper wetting action occurs.

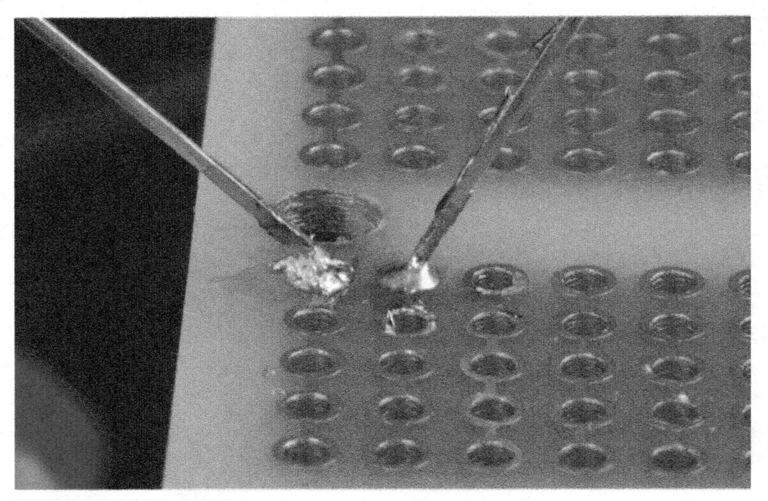

FIGURE 4-5: Joint disturbed when soldering

Wire Lead Soldered to PCB

Sometimes when you're adding a component to the PCB, you can bend the leads too far down and inadvertently solder the wire lead to the PCB (as shown in Figure 4-6). This will cause shorts and is generally a bad thing! You can easily avoid this issue by not bending the leads more than 45° from the connection point. To fix the problem, you are going to need to remove the solder from the entire connection and try again. Don't worry: removing solder isn't too difficult! And that's exactly what I want to talk about next.

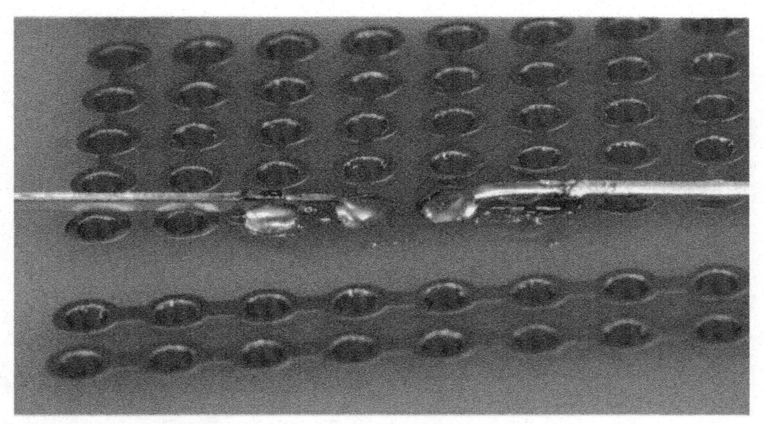

FIGURE 4-6: A wire lead lying flat and soldered to the PCB—not good!

As I mentioned earlier, the most common fix for many mistakes is simply to remove the solder and start over. Fortunately, removing solder is not difficult, and there are several ways of doing it. We'll examine a few of them in the next section, and learn about some helpful tools.

DIFFERENT WAYS TO REMOVE SOLDER

As I've mentioned throughout this chapter, the solution to several of the most common soldering problems is to remove solder. Depending on the issue, there are several different ways to go about it.

The easiest and fastest way to fix a solder bridge (see Figure 4-7) in a connection is to simply reheat the joint and wick some of it away. To do this, first make sure your iron is hot, and wipe the tip on a sponge to make sure it is clean of any excess. Next, reflow the joint and pull the tip of the iron away. (See Figure 4-8.) Wipe the solder off the tip by dabbing it on the sponge again. You can repeat this process a few times to see if it removes enough of the excess solder, but be careful not to overheat the solder or the PCB. When a solder bridge forms between two pads, most of the time a quick reheat solves the problem.

FIGURE 4-7: A solder bridge

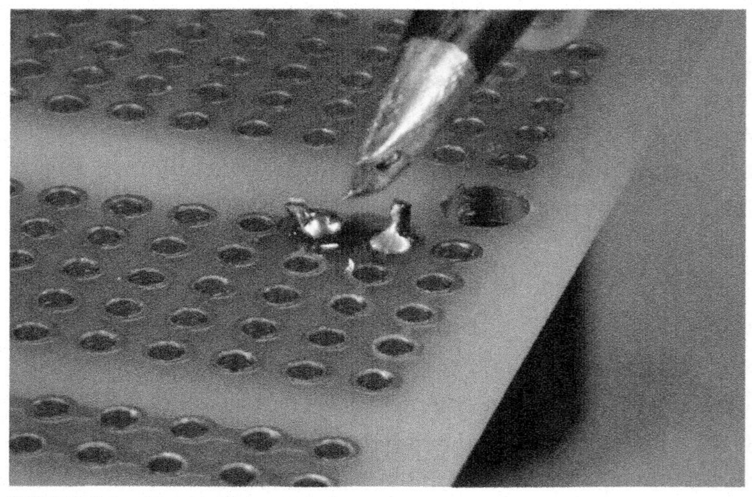

FIGURE 4-8: A solder bridge being repaired

When you have way too much solder and the simple reheat-and-wipe technique won't work, or if you have several solder bridges to deal with, the use of copper braid will easily solve the problem. Start by making sure your iron is clean and up to temperature. Next, place the copper braid on top of the solder that you want to remove. Place your iron on top of the copper braid and you will see the solder wick into the braid. (See Figure 4-9.) Once the braid looks saturated with solder, remove the iron and the braid at the same time. The key point here is *at the same time*. This ensures the still-molten solder will be removed along with the copper braid. This is a good solution for removing excess solder, but if you need to remove all the solder so you can remove a component, you need to use a solder sucker, and that's what we will look at next.

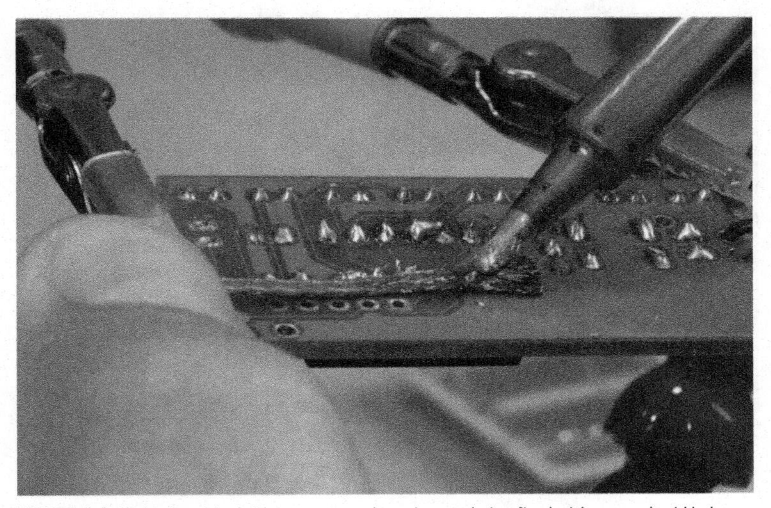

FIGURE 4-9: Multiple solder bridges on top and another set being fixed with copper braid below

> **NOTE** You will notice in Figure 4-9 that my soldering iron tip is getting oxidized quickly. Make sure that when using copper braid you continue to regularly clean the tip. This will make your iron work better, and will make the tip last a lot longer.

A *solder sucker* is a small, manual vacuum pump that will safely remove molten solder. Using a solder sucker is easy, but it requires excellent timing. Get the solder sucker ready to remove the solder by pressing down on the plunger. Next, with your soldering iron in one hand and the solder sucker in the other hand, heat the solder connection you would like to remove (see Figure 4-10). Now remove the iron, immediately place the solder sucker's tip on the solder connection, and hit the button. This creates a vacuum that sucks up the molten solder. Last, press down on the plunger and eject the cold solder into a proper container for disposing. Repeat, as necessary, to remove all the solder.

FIGURE 4-10: Component that needs the solder removed with a solder sucker

> **TIP** Removing solder requires split-second timing. I usually practice the motion at least once before I heat up the connection and use the solder sucker, especially with expensive components. A quick practice run is worth the extra few seconds.

The electric solder sucker, shown in Figure 4-11, is a new tool I found while doing research for this book, and after using it, I instantly fell in love with it as a teaching tool. Just like a standard solder sucker, you place the tip over the area that needs the solder removed. But this tool has an integrated heater, so you don't need to make the quick—and sometimes tricky—switch from heating with the iron to using the solder sucker. I don't think everyone needs this tool, since a standard solder sucker works fine. However, for a lab with multiple users, especially in a lab used to teach soldering, this can be a lifesaver (and a PCB saver!). For day-to-day use, I'll stick with a standard nonelectric version, but when I am teaching a group of students, I will certainly bring this tool with me to use for the inevitable mistakes.

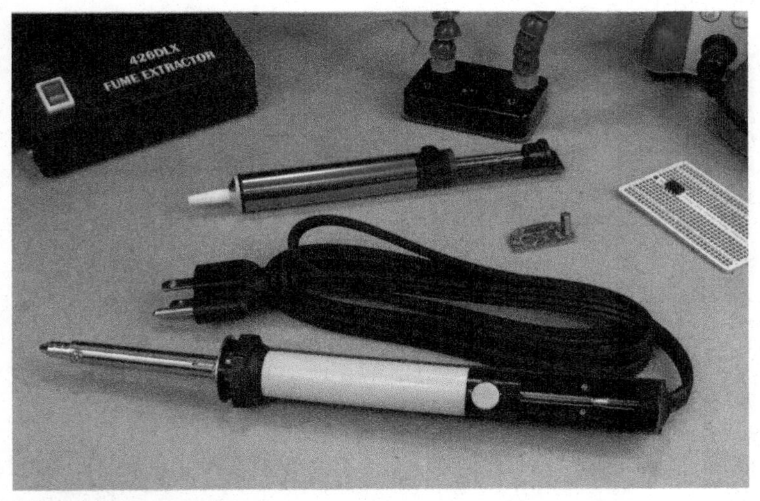

FIGURE 4-11: Electric solder sucker

After you use a solder sucker, you must eject the spent solder from the device. But where does that solder go? Typically, I see people eject it onto their work bench or, worse, the floor. If these little bits of solder get on your PCB, they can cause shorts. Worse, if the bits are lead-based they can be stepped on and tracked all over the place, causing an environmental hazard. Do yourself a favor and make a proper disposal system for your solder sucker—a solder sucker spittoon. (See Figure 4-12.) It's one of the easiest DIY projects for your electronics work bench. Simply find any flexible container and cut a small circle in the top. Once you use the solder sucker, place the tip of it through the opening in the top of the container and eject the spent solder. If you are using an electric solder sucker, though, make sure you cut a large enough opening so the heated tip does not melt the plastic.

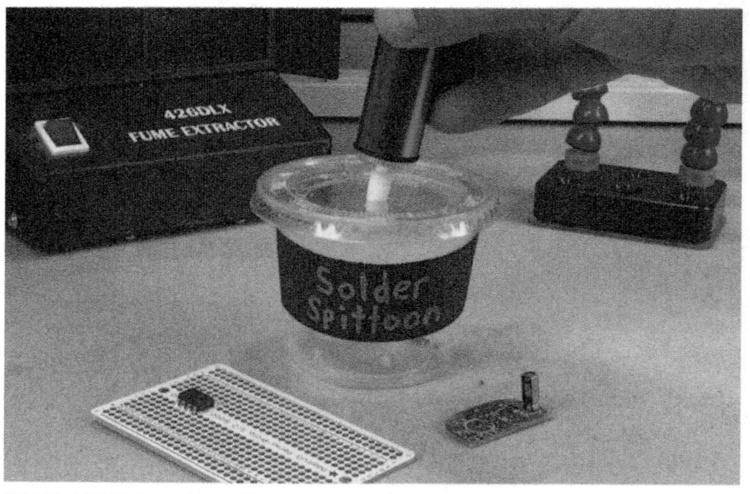

FIGURE 4-12: Make a solder sucker spittoon for safe disposal of your spent solder.

REPAIRING LIFTED PADS

Although it's rare, you may eventually need to repair a lifted pad on a PCB. This is typically caused by significantly overheating the PCB, but I have also seen this occur on poor-quality PCBs. If you purchase your parts from a reputable shop, it's almost never an issue. Poor-quality boards are usually very inexpensive and purchased from no-name online stores. I guess you get what you pay for.

To be totally honest, it was unbelievably difficult for me to make a solder pad lift on the high-quality PCBs I have in my studio. However, after heating and abusing the solder connection on the PCB for more than 30 seconds, which is about 26 seconds too long, I did manage to get a slightly lifted pad. It is amazing how difficult it is to make this type of mistake. It's a testament to buying your PCBs from well-known sources.

There are cases where you won't be able to fix the lifted pad, and sometimes a lifted pad won't affect the final project. However, most of the time you are going to want to try to fix the problem. Once you notice the pad is lifted, stop! Next, remove the solder using a solder sucker. Now, trim the component lead long enough to hold the pad in place, but not so long that it causes a short circuit with any other components. Next, bend the cut lead down against the PCB at a 90° angle and it will hold the pad in place. Last, solder the connection. In Figure 4-13, you can see that I bent the wire lead to the right, where there weren't any other components, and soldered it into place. This should fix the lifted pad. If it doesn't, you may be able to bypass the connection with a new piece of wire, or in the worst case you might have to start over with a new PCB.

Fortunately, a lifted pad is rare, and destroying a PCB beyond repair is exceptionally rare, especially after you have a few solder connections under your belt.

FIGURE 4-13: Repair to a pad that was damaged by extreme overheating of the PCB

5

Advanced Soldering

Up to this point, we have learned about soldering components that have two or more wire leads that are inserted into a printed circuit board (PCB). Now we are going to look at soldering *surface-mount devices (SMDs)*, which is soldering components that lie directly on the PCB and may or may not have any wire leads. This type of soldering is called *surface-mount technology (SMT)* and is typically done with more advanced tools, and most

often by a machine. But that doesn't mean you can't do it by hand, and that's what this chapter is all about.

There are a few reasons that you would want to use an SMD. The first would be that the component you want to use is only available in a surface mount package. Electronics are constantly shrinking in size, and although there are plenty of hobby electronics people out there that would love a larger form factor, sometimes it's just not an option. This mostly applies to higher-end sensors and integrated circuits (ICs), since there are still plenty of through-hole versions of LEDs, resistors, and ICs, with long leads that are easily soldered into a PCB. The other reason might be that you want to shrink your project down a little bit. I find myself using surface-mount LEDs all the time, since they are small and easy to solder. And to be honest, sometimes they just look cool, too. (See Figure 5-1.)

You might be asking yourself if you can really solder an SMD with a standard soldering iron. The answer is, yes—on one condition. The component you want to solder must have exposed leads. Parts without visible leads require a more complicated technique.

FIGURE 5-1: Soldering an SMD with a standard soldering iron.

Typically, they are soldered using either a hot air rework station or a reflow oven. (If you want to learn more about hot air rework stations, see the sidebar "Surface-Mount Soldering? Maybe a Hot Air Rework Station Is for You.") Those techniques are way beyond the scope of this book. Fortunately, there are a lot of components that have visible connection points, or leads that can easily be soldered by hand with a standard soldering iron.

There are a lot of things to learn when it comes to SMT, and I can't cover all the details in this book, like the naming conventions, numbering, and specific form factors in this book. However, I can show you the basics and give you some tips and tricks. I want you to learn something that took me far too long to figure out—you can easily solder many different kinds of SMDs with just a regular soldering iron!

TOOLS

I mentioned that you can solder surface-mount components with a standard soldering iron, and although that is completely true, there are a few extra tools and materials you might want to consider adding to your workbench to make handling the components a lot easier.

The most difficult part about learning to solder surface-mount components is the physical size of the components themselves—some SMDs make a grain of rice look big! That's why you are going to need some type of magnification to be able to solder the components, and to inspect the solder connections afterwards for shorts and any solder bridges formed from excess solder at the connection. In Figure 5-2, you can see the two different magnifying glasses that I use. I use the pair on the right with the grey strap for prolonged soldering sessions. They work better for me, but they aren't as comfortable as the black-framed version on the left. You can use a simple magnifying glass or jeweler's loupe, but those typically require an extra hand, and you're going to need both hands free to solder these small components.

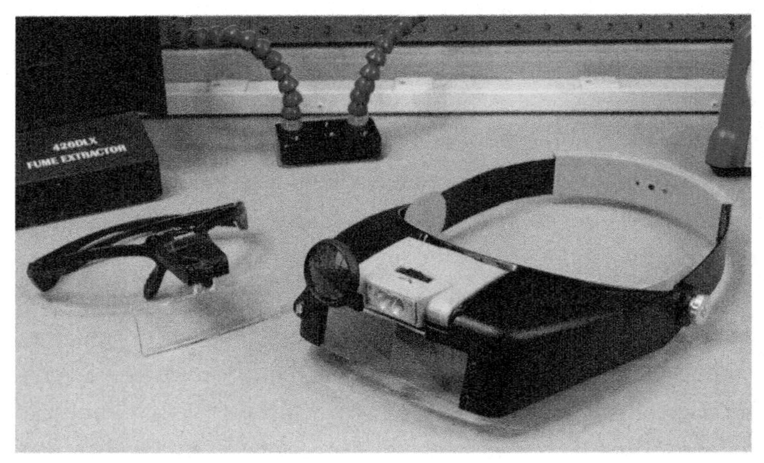

FIGURE 5-2: Magnifying glasses

In the past, I used a stereo microscope to solder surface-mount components, but after it was damaged beyond repair, I started to look for a more affordable alternative. Even a used stereo microscope will cost you hundreds of dollars, and I thought there had to be a better option. For a very short period of time I used a USB microscope, but although it was priced right at only about $30 USD, having to connect it to a computer was not always practical. Then, I discovered the self-contained digital microscope, pictured in Figure 5-3, and my soldering experience improved dramatically. It's incredibly easy to use, and has an internal rechargeable battery, dimmable lights, and a simple focusing system. You can even take pictures and store them on a micro SD card. What really sold me was the price—you can find them for about $60 USD online. Several different companies offered them, but they all had the same specifications and looked the same. I chose mine based on price and shipping, and it was well worth the investment. I use it all the time, and not just for soldering.

FIGURE 5-3: Self-contained digital magnifying microscope

As we discussed, handling small surface-mount technology (SMT) components can be challenging, so that's why you will need at least one pair of good tweezers, similar to the ones shown in Figure 5-4. These types of tweezers have tips that are either straight or curved, and both will work equally well. Make sure they are electrostatic discharge (ESD) safe and non-magnetic so they will not damage any of your components. Using a good pair of tweezers to place those tiny components will save you hours of frustration.

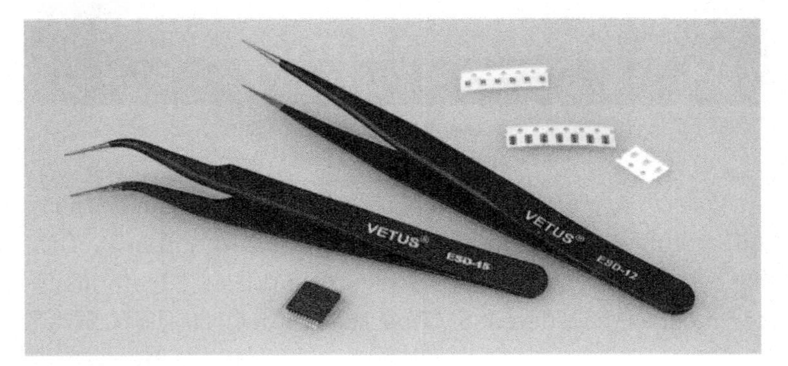

FIGURE 5-4: ESD-safe tweezers

When working with SMT you will most likely not use a third hand tool as much as you would when using standard through-hole components. Most people lay the board flat on their work surface, since there are not any wires going through the board. For this reason, you should invest in a high-quality silicone mat (shown in Figure 5-5). Not only will it protect your work surface from excessive heat, but it will also help keep all your components organized. If you plan on just using a soldering iron, a silicone mat isn't essential, but if you plan on doing any rework in the future, this is a must-have for your workbench. When selecting a mat, make sure it is specifically for soldering, as it needs to be able to handle high heat.

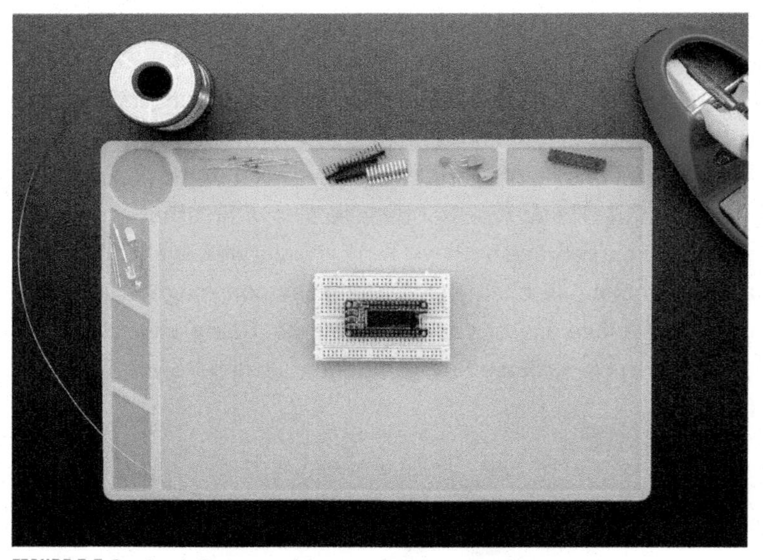

FIGURE 5-5: Insulated silicone rework mat IMAGE COURTESY OF ADAFRUIT INDUSTRIES, CC BY-NC-SA 2.0.

One of my favorite SMT tools is my "One PCB to Ruler Them All" from Adafruit Industries (see Figures 5-6 and 5-7). This handy little two-sided ruler has the form factors ("footprints") of almost all the different SMT packages, like QFN, TDFN, SOIC,

SOP, and more. I won't go into a lot of detail—that's for another book—but basically, it's the form factors of all the different manufactured components. It also has some other useful things, like wire gauge sizes, and of course it functions as a handy six-inch ruler. I use it all the time just to get an idea of how small or large a component will be.

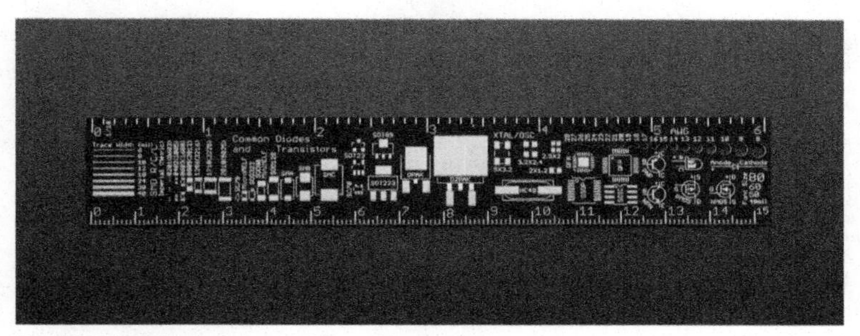

FIGURE 5-6: The front of the "One PCB to Ruler Them All"

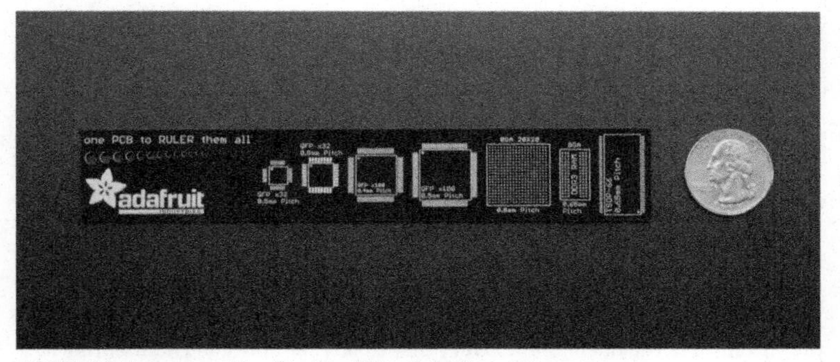

FIGURE 5-7: The reverse side of the "One PCB to Ruler Them All" IMAGE COURTESY OF ADAFRUIT INDUSTRIES, CC BY-NC-SA 2.0.

There are times it would be nice to have a little hand-held manual vacuum pump (shown in Figure 5-8) to place a larger SMD component on the PCB. If you have a few extra dollars to spend (they are literally only a few dollars), then pick one up and add it to your work bench. Otherwise, you can certainly make do with a nice pair of tweezers.

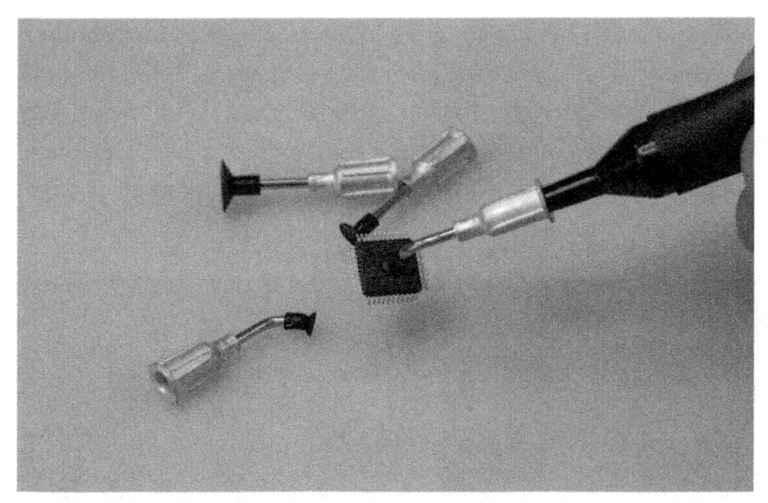

FIGURE 5-8: Hand-held vacuum pump

Surface-Mount Soldering? Maybe a Hot Air Rework Station Is for You.

Admittedly, you do not need a hot air rework station (shown in Figure 5-9) when you're learning to solder, even when using SMT, but I'm including it here because there are some benefits associated with having one in your lab. First, most come with a built-in, temperature-controlled soldering iron, which is what we are going to use. Second, you can use the hot air handle for other tasks, like shrinking heat-shrink tubing. Third, some even have a built-in fume-extraction system. These machines range in price, starting at about $100 USD. You will absolutely need one if you continue to expand your work with SMT, especially if your work involves high-end components that use ball grid array (BGA) technology. However, that is a bit beyond the scope of this book. OK, it's *way* beyond the scope of this book! When you get to that point in your soldering adventures, you'll know exactly when and why you need a hot air station.

FIGURE 5-9: Hot air rework station

MATERIALS

In addition to the materials you typically need for soldering, there are also a few materials that you will want to have on hand when you're learning to solder SMDs. You will use the same type of solder we discussed earlier in this book. Using a slightly thinner gauge is helpful, but it isn't necessary. Most of the other materials are directly related to the small size of the components, but some materials are necessary because of the specific way that you will be soldering SMDs. None of the materials are that expensive, and most can be used for other purposes in your lab.

In standard through-hole soldering, you don't necessarily need a flux pen, but when soldering SMDs you will need one to make a reliable connection. Using the flux pen (shown in Figure 5-10) in addition to flux-core solder will make the wetting action easier. Since the pads are small, even a little oxidation can make the soldering difficult. Also, the technique we will be using typically keeps the solder melted for a longer period of time, which will cause additional oxidation.

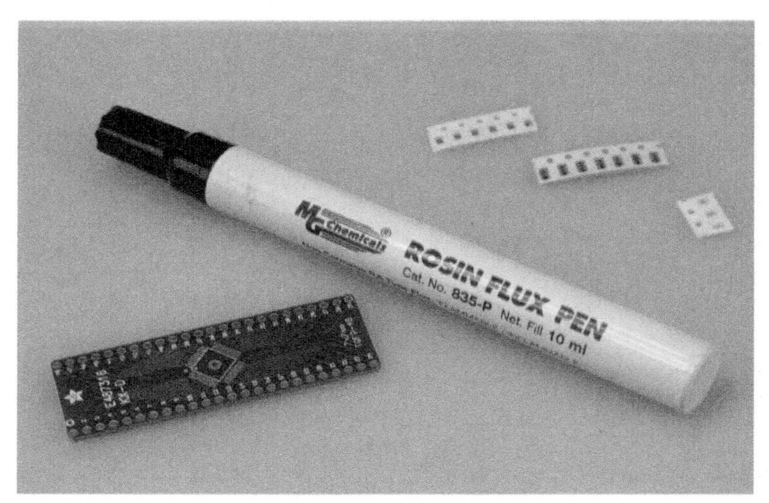

FIGURE 5-10: A flux pen is a must when surface-mount soldering.

The single most useful item to have when you're learning to solder SMD components is a breakout board. (See Figure 5-11.) These boards, available from several different companies, mean that you don't need to design your own board to be able to use a component that is only available in a surface-mount package. A breakout board will allow you to easily solder the component in place, and then you can solder wires "breaking out" the pins of the SMD. Be sure you know what type of SMT you are using and buy the matching breakout board—there are a lot of different form factors.

Chip Quik is a special flux and alloy that enables you to remove SMDs without damaging them. (See Figure 5-12.) Traditionally, you would have a hard time desoldering a component with a standard soldering iron. Traditionally, removing a part would require you to purchase a hot air reflow system. Chip Quik makes the whole process easy, and you only need a standard soldering iron. I'll cover the step-by-step process later in this chapter in the section, "When Things Go Wrong—Fixing and Removing Components."

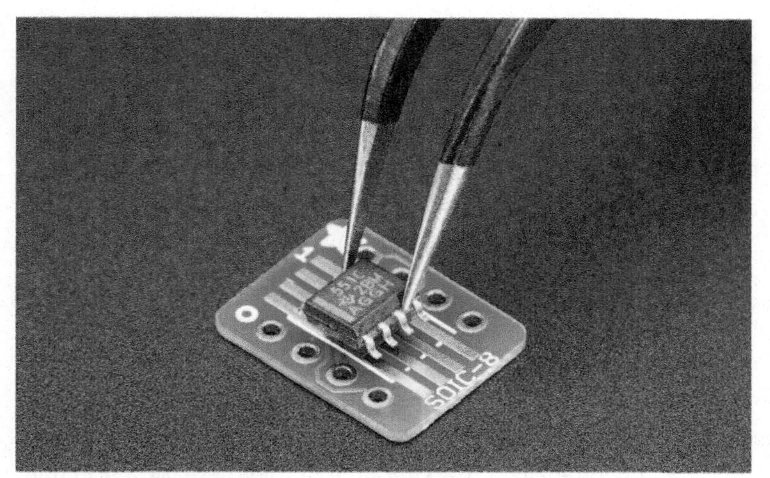

FIGURE 5-11: SMT breakout PCB for SOIC-8 IMAGE COURTESY OF ADAFRUIT INDUSTRIES, CC BY-NC-SA 2.0.

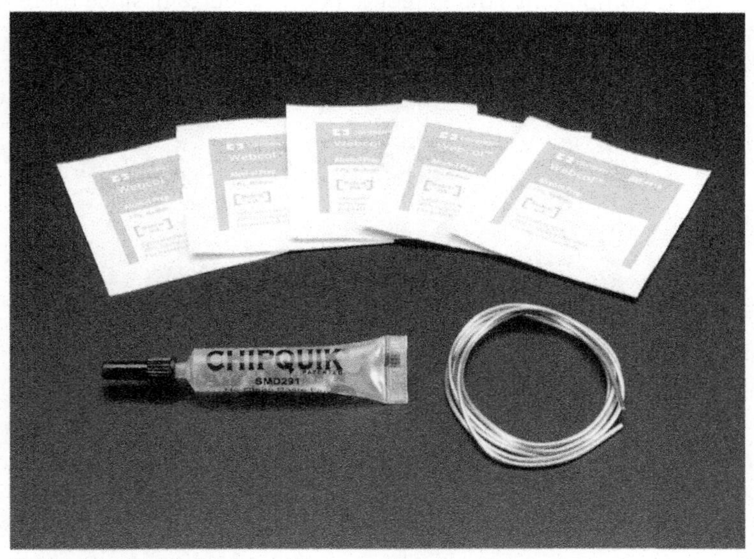

FIGURE 5-12: Chip Quik SMD Removal Kit IMAGE COURTESY OF ADAFRUIT INDUSTRIES, CC BY-NC-SA 2.0.

HOW TO SOLDER SIMPLE SURFACE MOUNT DEVICES

Soldering SMDs is a lot different than soldering through-hole parts. We'll start by soldering a 0805 LED, which doesn't have any visible leads, but is easily soldered in place. The term *0805* refers to the size—it's just about 0.08″ by 0.05″ long. It's small, but not even close to the size of a 0201 component! (See Figure 5-13.) The next step is to make sure you have the component oriented properly. A standard through-hole LED has a long lead and a short lead. The longer lead is the positive (+) lead. SMD manufacturers mark their components' polarity differently. On an LED, there is typically a marking on it somewhere, or a notch in the plastic that indicates the polarity, but you will need to reference the component's datasheet to know for certain, since there is a lot of variation between manufacturers.

FIGURE 5-13: The size of SMT components

First, start by wiping the flux pen on the pads of the PCB where you want to solder the component in place. This will help reduce any oxidation and improve the wetting action. Next, make sure your soldering iron is up to temperature and the tip is clean. Then melt a small amount of solder to the tip of your iron and place it on one of the pads where you plan to place the component and hold it there for 1–2 seconds. A small amount of solder should flow off the iron's tip and onto the pad of the PCB, as shown in Figure 5-14.

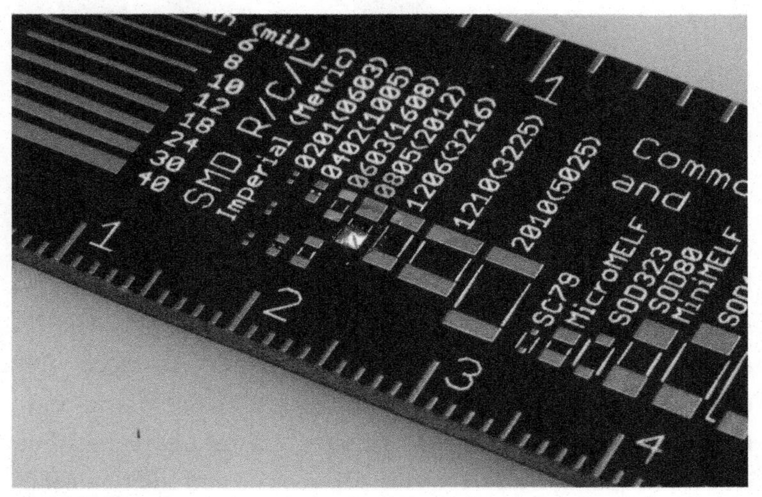

FIGURE 5-14: Adding solder to one of the pads, after using a flux pen, in preparation for soldering in an SMD LED

Next, place the component on the pre-tinned pad, and heat that connection up with your soldering iron. The solder will flow again and the component will be seated against the PCB. Remove the heat. This will hold the component in place for the next step. (See Figure 5-15.)

FIGURE 5-15: Soldering the surface-mount component in place

Now that one side is soldered, heat up the other side of the component and pad with your soldering iron. While they're being heated, add a small amount of solder to the connection point. A little solder goes a long way. It should end up looking similar to a through-hole component in terms of its general appearance, with a nice shiny solder connection that engulfs the end of the component. (See Figure 5-16.) Now, go back and make sure the other side is adequately soldered. If it's not, you can add additional flux and solder, as needed. In most cases, you won't have to add any solder, since you only need a small amount to secure SMD components. The last step is to clean up any excess flux with rubbing alcohol. I use prepackaged wipes that are specifically designed for this task.

HOW TO SOLDER A MULTI-LEAD SMD

Some SMT uses components with small wire leads, like those pictured in Figure 5-17. To solder this many leads, you could

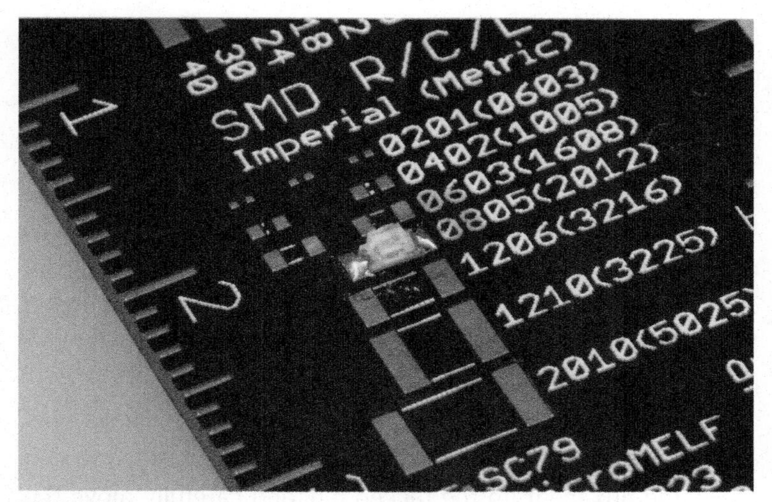

FIGURE 5-16: Soldering the surface-mount component in place

pre-tin the pads on the PCB, but the component may not end up in the proper place once all the solder flows. Fortunately, there is a much easier and more efficient way to make these types of connections.

FIGURE 5-17: Tacking the IC in place

Start by fluxing all the pads where the component will be soldered. Next, place the component on the PCB, aligning the pads to the pins. Make sure the component is oriented correctly; if you aren't sure, check the instructions or datasheet. Now, add a small amount of solder to the clean tip of your iron. Next, while holding the SMD in place, heat one of the legs of the component. Finally, remove the heat when the solder flows onto and under the leg of the SMD, which should only take 1–2 seconds, depending on the temperature of your iron. The component is now tacked into place, and you're ready for the next step.

Now that the component is tacked in place, make sure it's aligned properly with all the pads on the PCB. If not, you can add some flux, heat up the tacked pin, and carefully move the component into place.

Now, starting on a side of the SMD that is not tacked in place, apply solder and iron to one pin. As soon as the solder starts to melt, drag the tip and the solder across all the pins on that side of the component (as seen in Figure 5-18). This is called the *drag technique* for obvious reasons. Don't worry if there are solder bridges, since there will most likely be a few. It's more important to have enough solder and proper wetting of the pads and leads. Repeat this process on all sides of the SMD.

You will inevitably have excess solder on the leads of the components, and possibly a bridge or two. (See Figure 5-19.) The number of solder bridges you have greatly depends on how many leads there are and how closely they are spaced on the component. It may be just one or two of the leads, or it could be all of them. Don't panic; it's totally OK and easy to fix. In fact, it's better to have a little excess solder in this situation. You want to make sure there is a good wetting action on all the leads.

To fix any solder bridges and remove excess solder from the components, you will use some copper solder braid. Place the braid over all the leads on one side of the component while heating

FIGURE 5-18: Soldering ICs with the drag technique

FIGURE 5-19: Excess solder on the component, forming a solder bridge

it from above with your iron. This will wick up excess solder, leaving a nice clean connection. You can see on the left of Figure 5-20 that the leads have all been cleaned up. On the right, I'm in the middle of wicking up the excess. It only takes a few seconds of heat to remove the excess solder, and it makes the connections look great and, most important, work perfectly.

FIGURE 5-20: Excess solder being wicked up with copper solder braid

That's it! Simple; and you can see in Figure 5-21 that it looks like a professional fabrication house used their fancy equipment to solder that SMD in place. I wish I had known years ago just how easy it is to integrate SMT into my projects.

FIGURE 5-21: Finished SMD soldered in place

WHEN SOMETHING GOES WRONG—FIXING AND REMOVING COMPONENTS

You might be asking yourself, "What happens if I make a mistake? It's hard enough to solder the components in place. How can I remove multi-lead SMD components with just a soldering iron?" It's not as difficult as it sounds, and you don't need a hot air rework station—just a little Chip Quik and a regular soldering iron. Chip Quik uses a special metal alloy that will bond to your solder and keep it flowing longer (see Figure 5-22). This will allow you to heat up all the leads with just a standard iron, and then pluck the component away from the pad with tweezers. Start by applying the Chip Quik flux to all the leads of the component, as shown in Figure 5-22. Next, let the special metal alloy flow onto all the leads you want to remove, as pictured in Figure 5-23. Next, go back and heat all the leads again with your iron. Remember: the Chip Quik alloy will stay molten for a longer period of time than standard solder. Now, quickly grab the component with tweezers and remove it from the PCB. (See Figure 5-24.)

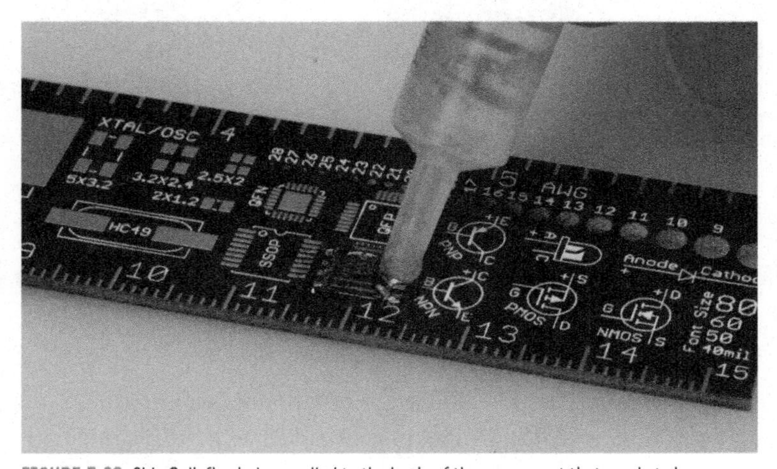

FIGURE 5-22: Chip Quik flux being applied to the leads of the component that needs to be removed

FIGURE 5-23: Chip Quik alloy being used on a PCB to remove a component

FIGURE 5-24: The component successfully removed

The last step is to clean up the remaining solder with some copper solder braid, and wipe down the PCB with the alcohol pad supplied in the Chip Quik kit. This is a great solution for removing components without the need for a hot air rework station.

SUMMARY

We have covered the absolute basics of soldering surface-mount components, but there is so much more to cover, like stencils, reflow ovens, and more. It's enough for another whole book—or even two! Hopefully, this book got you started in the right direction and made you want to learn more. Happy soldering!

Index

T

U

V

W

CPSIA information can be obtained
at www.ICGtesting.com
Printed in the USA
BVOW10s1517121017
497431BV00004B/71/P

9 781680 453843